CHEMISTRY OF NUCLEAR POWER

CHEMISTRY OF NUCLEAR POWER

By

John Keith

J. K. DAWSON, Ph.D., F.R.I.C.

Senior Principal Scientist,
Reactor Chemistry Group, A.E.R.E., Harwell

and

G. LONG, Ph.D., A.R.I.C.

Reactor Chemistry Group, A.E.R.E., Harwell

PHILOSOPHICAL LIBRARY INC.

15 EAST FORTIETH STREET

NEW YORK 16, N.Y.

First published 1959

Published, 1959, by Philosophical Library Inc.
15 East 40th Street, New York 16, N.Y.

Printed in Great Britain for Philosophical Library by
The Whitefriars Press Ltd., London and Tonbridge

PREFACE

FISSION was discovered in 1939, the first atomic bomb was exploded over Hiroshima in 1945, and the world's first large-scale atomic power station for electricity production was opened by H.M. the Queen at Calder Hall in 1956. In this period of under twenty years we have witnessed the growth of a few painstaking laboratory experiments into a major industry. These first experiments were performed by a chemist, but the spectacular results have been achieved by the combined efforts of scientists and engineers of many kinds. It is inevitable that the part played by the engineers should have received widespread recognition, since it is they who have been ultimately responsible for the design of such highly complex installations as the experimental fast reactor at Dounreay, atomic submarine engines for the American *Nautilus* and *Skate*, and many other reactors. Chemists, however, have also had a vital part to play, and it is the purpose of this book to describe their contribution to the overall programme.

We have not attempted to give a minutely detailed account; the reader is referred to the list of key references at the end of each chapter if he wishes to develop a more intimate knowledge of this field. Instead, we have tried to give a broad picture of established practice and present lines of thought, with some indication of the problems requiring future effort.

In such a complex industry it is inevitable that fields of research should overlap, and it is often impossible to assign a given topic to any one particular scientific discipline. Furthermore, in order to understand fully the nature of the problems facing the chemist it is necessary to have some knowledge of the non-chemical background of the problem. We have therefore found it necessary to make an occasional digression from the main topic of chemistry into the realms of metallurgy and physics. We trust that this book will be of use to those who have had some scientific training and who wish to become more closely acquainted with the impact of atomic power development upon almost all of the elements in the Periodic Table.

The authors are indebted in particular to Dr. R. Spence, Chief Chemist, A.E.R.E., and to Dr. R. G. Sowden who made valuable comments on the manuscript, and to many members of the staff of

A.E.R.E., Harwell, for their willing co-operation. They also wish to acknowledge the permission of the U.K.A.E.A. to use material contained in unclassified A.E.R.E. Reports.

July, 1958   J. K. DAWSON
                                           G. LONG

Footnote

Since the manuscript of this book was written, some 2,500 papers have appeared for the Second United Nations Conference on the Peaceful Uses of Atomic Energy held in September, 1958. Much detailed information on topics covered by this book was presented at this Conference, but generally the basic chemical principles remained unchanged and extensive modifications to the text were not required. However, a few of the papers presented entirely new information, and the salient features of these have been incorporated. The printed Proceedings are not yet available, but important individual papers have been included in the lists for further reading.

November, 1958

CONTENTS

ERRATA

Page 53 (line 36)   *for* 15.10^{18} *read* 15.10^{16}

Page 56 (formula)   *for* $2.1\ 10^6\text{y}$ *read* $2 \cdot 1.10^6\text{y}$

                      for 3.210^6 barns   *read* $3 \cdot 2.10^6$ barns

Page 162 (line 32)   *for* $U(VI)$ *read* $U(IV)$

PLATES

(Between pages 96 *and* 97)

The authors and publishers wish to acknowledge the permission given by the United Kingdom Atomic Energy Authority to reproduce these photographs, and the valuable assistance given by members of the Metallurgy Division, Harwell, in compiling Plate VIII.

CHAPTER I

THE ROLE OF THE CHEMIST IN THE ATOMIC ENERGY INDUSTRY

THE GROWTH OF THE INDUSTRY

THE growth of the atomic energy industry has been unique. At no other time in the history of scientific development have a few comparatively simple observations by a handful of scientists provoked such immediate world-wide repercussions. In less than a decade weapons of unparalleled destructive power were developed, and after a further decade the large-scale production of useful power from nuclear fission has become a practical reality. Such an achievement has called for the closest collaboration between men of nearly all the scientific disciplines; at all stages of development the chemist has had a vital contribution to make.

THE BEGINNINGS

No sooner had the process of fission of U^{235} by neutrons been demonstrated early in 1939 than scientists in Britain and the United States appreciated the potentialities of the process as a weapon, and made suitable representations to their Governments. By mid-1941, a small Anglo-French team had established the feasibility of constructing both a controlled chain-reacting pile and a weapon. At that time British Industry was strained to capacity by the war, and it was not possible to set up the necessary organisation for the development and production of a nuclear weapon. The responsibility for the development of a weapon rested therefore with the United States. Two groups of British scientists, working in the United States, contributed to the successful development of the weapon. In 1943 a joint Anglo-Canadian project was also set up to build the heavy-water moderated reactor at Chalk River. This team was later to lay the foundations of the post-war British project.

In the years 1942 to 1945 the American teams made great efforts to obtain the necessary basic information and to build up the complex industrial plant for producing the fissile materials U^{235} and Pu^{239} essential for a nuclear weapon. In August 1945 the objects of the project were accomplished when nuclear devices were detonated over the cities of Hiroshima and Nagasaki.

At the end of the war the British teams were withdrawn from the United States and plans were drawn up for the British atomic energy programme. It was originally intended that effort should be directed towards the production of the necessary materials for weapons and power reactors, the development of the peaceful uses of atomic energy and the furtherance of fundamental research. In 1946, however, the *MacMahon* Atomic Energy Act was passed by Congress. This Act prevented collaboration with other countries, and it was therefore essential that work on the development of a British nuclear weapon should be initiated without delay. The years 1947–48 therefore saw a repetition of the rapid expansion which had occurred in the United States in the early 1940's. Under the auspices of the Ministry of Supply, two Research Centres were established in Berkshire; one, at Harwell, for fundamental research, and the other, at Aldermaston, for weapon development. The headquarters of the Industrial Group was set up at Risley in Lancashire; this Group was responsible for the rapid construction and operation of three factories for the production of uranium and plutonium. In October 1952 the first all-British nuclear weapon was successfully detonated on the island of Monte Bello, off the coast of Australia.

PEACEFUL USES

During these five years of intensive expansion for weapon development the peaceful applications of atomic energy were not neglected. It was becoming increasingly apparent that the development of some new source of power was vital to Britain's prosperity. Not only could a diminution of reserves of coal be anticipated in the foreseeable future, but the situation was aggravated by the growing demands for power. In an expanding industrial economy, with its increasing mechanisation, Britain's power requirements have been approximately doubling every eleven years for several decades. Other indigenous sources of power, such as oil shale or water power, are of only limited availability, and in recent years it has become increasingly apparent that too great a dependence upon supplies of imported fuel oils is politically undesirable. Nuclear fission offered the very real possibility of meeting the extension of our power requirements for many years. The nuclear fuels— uranium and thorium—are available in considerable quantities in several areas of the world, including parts of the Commonwealth.

So it was that in the late 1940's an extensive programme for the development of atomic power was initiated, which culminated in the successful commissioning, in October 1956, of the world's first large-scale nuclear power-station at Calder Hall. In a decade the industry had developed into a nation-wide organisation, involving a capital investment of hundreds of millions of pounds, employing some tens of thousands of men and providing, indirectly, employment for many more.

The development of such a vast project *ab initio* calls for a massive research effort, but any such effort is completely impotent without the sound backing of several vital industries, mainly heavy, civil and electronic engineering. Extremely high standards are expected of these supporting industries, and extensive development work is often required.

Not only is a highly organised industry necessary for the successful development of nuclear power; a high level of capital investment is also required. These two factors—the cost and the necessary engineering backing—has confined the initial exploitation of nuclear power to the highly industrialised nations, namely Britain, Canada, the United States, the Soviet Union and France. Information on the basic technology is now becoming more generally available, and more countries are entering the field of nuclear development. These include India, Germany, Sweden and several European countries collectively under the Euratom and O.E.E.C. organisations. Some of these countries are, however, finding it necessary to build up the engineering background which is so essential to the successful development of a nuclear power industry.

THE UNITED KINGDOM ATOMIC ENERGY AUTHORITY

As the emphasis on the development of nuclear power increased it became apparent that the organisation within the Ministry of Supply was administratively unsuitable. The atomic energy project was becoming one of Britain's major industries, and called for an organisation working on industrial lines. Accordingly, in January 1954 the United Kingdom Atomic Energy Authority was set up by Act of Parliament under the auspices of the Lord President of the Council. The functions of the Authority may be divided into Research and Production.

Research

Fundamental work on the many aspects of nuclear power is primarily carried out at the Research Establishment at Harwell, and work is also undertaken on contract by Universities and

Industrial Research Laboratories. The Industrial Group, with
headquarters at Risley, is responsible for the exploitation of any
development arising from the work at Harwell, taking charge of
the design and carrying out such applied research as may be
necessary. The Group has research laboratories at each of the
factories; in these the day-to-day problems which arise are met,
and work on the development of existing processes is carried out.
Problems of a more fundamental nature are referred back to the
Research Group at Harwell.

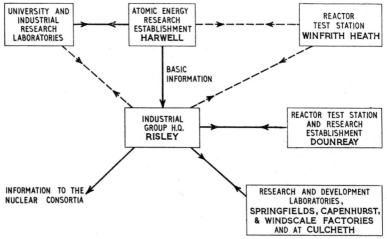

Fig. 1.   Research in the U.K.A.E.A.

Reactor research centres, at which advanced experimental
reactors are to be built, have been established, one by the Industrial
Group at Dounreay in Caithness and the other by the Research
Group at Winfrith Heath in Dorset.

The relationship between the various research organisations is
depicted in Fig. 1.

Production

The Industrial Group operates three large factories, located in
the North-West of England.

At Springfields, near Preston, uranium ore concentrates are
processed and converted to either the metal for reactor fuel elements
or the hexafluoride for the factory at Capenhurst, near Chester.
Here $U^{235}F_6$ is separated from the hexafluoride of natural uranium
by gaseous diffusion. Some of the enriched uranium is returned

to Springfields for the replenishment of the uranium recovered from irradiated fuel elements and some is sent to Dounreay for fabrication into fuel elements for the research reactors, such as DIDO and PLUTO at Harwell and the experimental fast reactor at Dounreay. At Windscale, natural uranium has been irradiated for plutonium production in the two reactors now shut down as a result of the accident on October 10th, 1957. Further irradiations are being continued in the four reactors at neighbouring Calder Hall, with the simultaneous production of electricity. The plutonium is separated chemically from the parent uranium and the fission products at the adjoining separation plant. The recovered uranium is returned to Springfields for refabrication into fuel elements. The Windscale processing plant, originally built to process the fuel from the two Windscale plutonium production piles, has sufficient additional capacity to carry out the reprocessing of the fuel elements from the U.K.A.E.A. reactors now in operation or under construction; a new plant is to be built to process the fuel elements from the civil reactors which are to be completed in the next few years.

The relation between the Industrial Group factories is shown in Fig. 2.

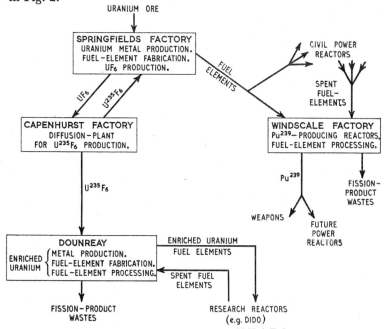

FIG. 2. PRODUCTION IN THE U.K.A.E.A.

THE IMPACT OF NUCLEAR POWER ON CHEMISTRY

STIMULATION OF INORGANIC CHEMISTRY

As a consequence of the exciting developments which took place in physical and organic chemistry in the first part of the twentieth century the study of pure inorganic chemistry fell into decline. In the early 1940's the pressing needs of the atomic energy programme stimulated a vast amount of new research in the inorganic field.

The chemist is called upon to help in the development of the technology of elements which, prior to 1940, were not widely used. These include zirconium and beryllium and even the nuclear raw materials themselves—uranium and thorium.   Previously these latter elements were mainly used, as oxides, in the pigmentation of glass and glazes and in the manufacture of Welsbach incandescent gas mantles.   The lack of fundamental knowledge of these materials is illustrated by the quoted melting point of uranium metal.   As late as 1943 this was given rather vaguely as " less than 1850°C "; it is in fact 1132°C!

In the production of U^{235}, the volatile uranium hexafluoride was required in large quantities for the diffusion plants; this necessitated an extensive development in the chemistry of fluorine and fluorides generally.   Interesting by-products of this research are the poly-fluorohydrocarbons P.T.F.E. (polytetrafluoroethylene) and *Kel-F* (polychlorotrifluoroethylene).   Both of these materials now find a wide application in the chemical industry as corrosion-resistant containers.

In general, there are very few elements which do not hold some interest for the worker in the field of atomic energy.   The lighter elements (H, Be, C) are suitable neutron moderators; most of the elements near the centre of the Periodic Table appear as fission products; the heavier elements are either potential nuclear fuels or are produced by neutron capture in the fuel during operation of a reactor.   Many of the transition metals make suitable constructional or container materials for nuclear reactors, and many elements, such as fluorine, as UF_6, find a unique application in the industry.

Besides being called upon to develop the chemistry of less known elements, the chemist has been presented with some new elements. Polonium and actinium, which are available naturally in only small quantities, can be produced by neutron capture on the irradiation of bismuth and radium respectively in a nuclear reactor. The transuranium elements—numbering at present ten in all—are produced either by neutron irradiation or heavy ion bombardment of uranium and other heavy elements. The chemistry of one of the transuranium elements—plutonium—is of vital interest to the atomic energy industry as this element constitutes an important nuclear fuel. Gaps in the Periodic Table have also been filled by the isolation of the elements technetium (43) and promethium (51) from the fission products, and by the production of astatine (85) by the irradiation of Bi^{210} with high-energy alpha particles from an accelerator.

SEPARATION PROCESSES

An interesting chemical aspect arises in the separation of isotopes, such as deuterium from hydrogen, and of closely related materials, such as zirconium and hafnium which occur together in nature. Hafnium, because of its high absorption of neutrons, is an undesirable contaminant of the important reactor constructional material zirconium. The chemical properties of the two elements are so alike that it is not possible to effect an efficient separation by classical chemical methods. Separation is now achieved by processes of either ion exchange or solvent extraction.

The outstanding example of isotope separation is that of separating U^{235} from natural uranium, in which it occurs to an extent of only 0·7 per cent. The separation process relies upon the slight difference in the rate of diffusion through a membrane of the hexafluorides $U^{235}F_6$ and $U^{238}F_6$. Fortunately the situation is not complicated by varying isotopic weights of fluorine, since only one isotope (F^{19}) occurs naturally. Although this separation is a physical process, considerable chemical effort was required before any plant could be operated. An understanding was needed of the various methods of preparing uranium hexafluoride, its physical properties and, of particular importance, its rate of attack on possible plant constructional materials.

SOME NOVEL TECHNIQUES

In building up the atomic energy industry many techniques novel to inorganic chemistry were developed and applied on an industrial scale.

Solvent extraction is a well-established technique of organic chemistry. Applied to inorganic materials it has found an important use in the processing of spent fuel elements. Uranyl and plutonyl nitrates react with certain organic reagents, such as tributyl phosphate, to form complexes of the type

$$UO_2(NO_3)_2[(RO)_3PO_4]_2$$

These complexes, presenting an organic structure to a solvent, are more soluble in non-polar solvents than in water, and so are extracted from aqueous solution. Other species, such as fission products and plutonium in the tetravalent state, do not form complexes and are therefore not extracted.

Inorganic ion-exchange materials have been used on an industrial scale for some time for the softening of water. Similar materials, such as certain clays, are at present being investigated as possible media on which to deposit fission-product cations prior to disposal. For more refined purposes, such as laboratory work or the deionisation of the cooling water for a nuclear reactor, synthetic ion-exchange materials can be tailor-made to specification. By incorporating into organic polymers suitable reactive groups, such as $-SO_3H$ or $-NR_2$, resins having cationic or anionic exchange properties can be produced.

A column containing a bed of fine particles of resin can be used to effect the separation of a mixture of ions. This can be achieved by, for example, forming a suitable anionic complex of the cation concerned, and absorbing this on an anionic exchange column. Cations which do not form anionic complexes pass through unabsorbed. Alternatively, a series of closely related elements may be separated by absorption on a cation exchange column followed by elution with a solution of a suitable complexing agent. Because of the slightly differing degrees of absorption on the ion-exchange resin and the varying strengths of complex formation, the different components are eluted successively from the column and may be collected separately. Such a technique is widely used for the rapid separation of mixtures of transuranium elements on a very small scale.

MICRO TECHNIQUES

The development of atomic energy, particularly in the early stages, required accurate measurements to be made on very small amounts of material. For example, at Chalk River the chemistry of plutonium was studied using no more than a few milligrams of the element, and the flow-sheet for the Windscale processing plant was based on these investigations.

To weigh the small quantities of material involved a quartz-fibre micro-balance is employed. The normal knife-edge of a balance is replaced by a fine, taut, horizontal thread of silica, which is firmly affixed to the mid-point of a light-weight beam. The turning moment of the sample on the pan is counteracted by a torque in the silica thread, applied by means of a rotating scale.

The balance point is indicated by an optical lever. Such a balance can be of very high sensitivity—10^{-9}g—but is limited to a maximum loading of a few mg by the strength of torsion thread. In an alternative design, this difficulty is overcome by supporting the beam and sample at the mid-point of the beam by a fine vertical silica thread, the torsion fibre still being retained as a means of balancing the beam. With such a system samples up to 5 g in weight can be accommodated without incurring any loss of sensitivity.

It is often desirable to carry out calorimetric measurements using small samples of material. A microcalorimeter has been developed at Harwell for the rapid estimation of small amounts of polonium. The method is based on a measurement of the heat liberated by the alpha decay of polonium, which amounts to 0.14 W/mg of Po^{210}. The sample, usually mounted on a foil or in an X-ray capillary, is placed in a small block of aluminium, the temperature of which is measured by means of a thermistor. The voltage across the thermistor is applied to a large-capacity condenser and the charging current is measured on a sensitive galvanometer. The observed charging current is proportional to the rate of change of e.m.f. across the thermistor, which, over the small temperature ranges observed, is proportional to the rate of change of the calorimeter temperature. Over the initial stages this is constant, and is directly proportional to the heat output of the sample. Thus, the quantity of polonium can be estimated by the observation of an almost constant galvanometer reading. The time taken for the measurement is about twenty minutes, compared with a matter of hours which is required to carry out microcalorimetric measurements by more conventional methods. By suitable design of the apparatus power inputs as low as 0.01 W can be accurately determined. This corresponds to only 70 μg of polonium.

The technique of polarography is well-established as a means of determining the concentrations of a mixture of cations in dilute aqueous solution, but the method lacks discrimination if two cations being determined have similar reduction potentials or if one is in large excess. A modified apparatus has been developed at Harwell and is now available commercially. In this instrument a small square-wave voltage, of the order of tens of mV, is superimposed upon the normal slowly-changing potential applied to the dropping electrode. The a.c. component of the current flowing is suitably filtered, amplified and displayed on a pen recorder. This technique not only eliminates the undesirable effects of the double-layer capacity current, but also in effect measures the derivative of the current-voltage curve. The resulting polarogram consists, therefore, of a series of peaks occurring at voltages which are characteristics of each cation present. At concentrations as low as 2.10^{-5} M the peaks corresponding to the reduction of Cu^{++} ($\sim 0.25V$) Pb^{++} ($\sim 0.47V$) Tl^{+} ($\sim 0.52V$), In^{+++} ($\sim 0.64V$) Cd^{++} ($\sim 0.68V$) and Zn^{++} ($\sim 1.08V$) can be resolved, and the resolution is not impaired even in the presence of a hundred-fold excess of cupric ions.

NEW TECHNIQUES MADE AVAILABLE BY THE OPERATION OF NUCLEAR REACTORS

Besides prompting the extensive development of several novel chemical techniques the atomic energy industry has made its own unique contributions to the science of chemistry. These contributions arise largely as a result of the operation of research reactors.

ISOTOPIC TRACERS

The irradiation in a nuclear reactor of many naturally occurring elements results in the production of radioactive isotopes. These find a wide variety of uses in all branches of chemistry. For example, compounds labelled with C^{14} are used in organic chemistry, and particularly biochemistry, in elucidating reaction mechanisms. Flourishing isotope production branches, including the Isotope Division at Harwell and the Radiochemical Centre at Amersham, have grown up. From these centres, radioactive isotopes are shipped in ever increasing amounts to all parts of the world. The catalogues list over 100 radioisotopes which are available, besides some 150 organic compounds labelled with C^{14}. Unfortunately the interesting story of the application of radioisotopes is beyond the scope of this book and will not be discussed further.

ACTIVATION ANALYSIS

The same process of activation of elements in a nuclear reactor forms the basis of an extremely sensitive method of determining a wide range of elements. As an example we may take the method of estimating arsenic, for which this technique has been used to determine trace amounts of the element both in biological systems and in germanium in the development of transistors.

The sample, containing as little as 10^{-10} g of arsenic, is irradiated in a nuclear reactor alongside a standard sample containing a known amount of arsenic. After irradiation both the standard and sample are, if necessary, dissolved and a known large excess of

THE ROLE OF THE CHEMIST 11

inactive arsenic added to act as a carrier for the minute amount of radioactive material produced in the reactor irradiation. The solutions are then purified from all other radioactive elements which may be present by suitable chemical techniques such as precipitation, distillation, chromatography and ion-exchange. Finally the sample is mounted on a counting tray and estimated by beta or gamma counting, and the quantity of arsenic in the sample obtained by comparison with the activity of the standard. The high sensitivity which is obtained in activation analysis arises mainly from the very small activity required for beta counting. With standard equipment 100 disintegrations per minute can be measured with reasonable precision. The sensitivity of the method, however, varies considerably from element to element, since the counting rate obtained for a given weight of sample depends upon the degree of neutron absorption of the element and upon the half-life for the decay of the active product. The weights of various elements which can be estimated with a precision of 10 per cent are shown in Table 1. These apply to activations carried out in the BEPO reactor. In a reactor of higher flux, such as DIDO, greater activities are produced and the sensitivity is increased about fifty-fold.

TABLE 1.—ESTIMATED SENSITIVITY OF ANALYSIS OF SELECTED ELEMENTS BY RADIOACTIVATION METHODS USING THE BEPO REACTOR

Element	Estimated Sensitivity (g)
Arsenic	5.10^{-11}
Copper	1.10^{-10}
Gold	5.10^{-2}
Iron	1.10^{-7}
Lead	5.10^{-6}
Sodium	1.10^{-10}
Sulphur	5.10^{-11}
Uranium	1.10^{-10}

RADIATION CHEMISTRY

Prior to 1940 the only sources of nuclear radiation were X-ray tubes, comparatively weak naturally-occurring radioactive materials (e.g. radium) and particle accelerators (e.g. the Van de Graaff generator). The operation of nuclear reactors enables a wide variety of radiation sources to be prepared by the neutron irradiation of suitable materials. Typical examples are listed in Table 2, in which radium is included for comparison.

It will be seen that some of the artificially-produced radio-isotopes are of much greater activity than the naturally-occurring radium, and furthermore they are available in considerably larger quantities.

TABLE 2.—RADIATION SOURCES PRODUCED BY NEUTRON
IRRADIATON IN A NUCLEAR REACTOR

Material Irradiated	Product	Principal Mode of Decay, and Energy (MeV)	Half-life for Decay	Approximate Strength of Source (curies/g)
Naturally occurring	Radium + daughters	α, 5 γ, 0·2	1 590 y	1·0
Cobalt	Co⁶⁰	γ, 1·3	5·2 y	up to 10
Uranium 235	Fission products, including:—			
	⎰ Cs¹³⁷ ⎱ Ba¹³⁷	β, 1·18⎱ γ, 0·6 ⎰	30 y	60
	Sr⁹⁰	β, 0·5	28 y	64
Uranium 238	Pu²³⁹	α, 5·1	24 300 y	0·07
	Am²⁴¹	α, 5·5	470 y	3·8
Bismuth	Po²¹⁰	α, 5·0	138 d	4700

The nuclear reactor itself is a large, intense source of radiation, consisting primarily of neutron and gamma radiation. The use of a reactor as a radiation source is hindered by the necessary presence of a flux of slow neutrons, which can give rise to intense radioactivity in the irradiated materials by neutron activation. This radioactivity can hamper the subsequent handling operations. However, although high gamma doses can be obtained in, for example, a Co⁶⁰ source, a nuclear reactor is the only source of a high flux of fast neutrons.

By incorporating a quantity of certain materials into the irradiated sample the nature of the irradiation can be completely changed and the amount of energy deposited within the sample increased by as much as several orders of magnitude. With an impregnation of U²³⁵, for example, fission occurs in the neutron flux of the reactor and the sample is effectively irradiated by fission fragments.   In BEPO the rate of energy deposition in water is increased by a factor of 10⁵ when uranyl sulphate is dissolved in the water to give a concentration of 1 g U²³⁵ per litre. Alternatively, incorporation of lithium or boron results in irradiation by high-energy alpha particles arising from the reactions

$$B^{10} + n \rightarrow Li^7 + \alpha \quad \text{and} \quad Li^6 + n \rightarrow H^3 + \alpha$$

THE CHEMIST'S CONTRIBUTION TO THE DEVELOPMENT OF ATOMIC ENERGY

Analytical chemistry is of vital importance at all of the atomic energy factories for the purposes of the control of purity and the prevention of excessive losses of valuable material in the plant effluents. In addition, at Capenhurst and Windscale potentially dangerous fissile materials are processed. Here the plants are in the main constructed in such a way that no combination of circumstances could give rise to an inadvertent nuclear chain reaction, with resulting damage to plant and injury to personnel. A most careful analytical accounting is still necessary, however, to ensure that excessive amounts of fissile material do not accumulate within the plant, for example as sludges.

The chemist has an important contribution to make in the research involved in the development of a reactor project and the associated processes. The fields of research fall broadly into five categories.

(i) Reactor Materials

A power-producing reactor requires between tens and hundreds of tons of highly purified materials, the exact nature of which depends upon the reactor type. For example, the Calder Hall reactors require uranium for the fuel, graphite for the moderator and carbon dioxide for the coolant. In some circumstances considerable quantities are required of materials, such as zirconium or beryllium, which until a few years ago were little known outside the laboratory. This has involved the development of processes for the refining of the ores and the reduction to the metals.

(ii) Reactor Development

The first reactors to be constructed are operated at low temperatures (below 200°C) and present comparatively few chemical problems. The higher temperatures which now occur in power-producing reactors mean that chemical reaction rates are very much increased, and problems of the compatibility of fuel, moderator and coolant arise. These problems are often aggravated by the presence of the intense radiation field within the reactor core. The development of novel high-temperature reactor systems, involving liquid metals, fused salts or aqueous solutions of uranium also presents formidable chemical problems, mainly of compatibility and radiation decomposition. Furthermore, to be economic a reactor must operate for many years, so that even very slow reactions (for example corrosion reactions) are therefore of importance in the successful running of a reactor.

In contrast to more conventional corrosion studies it is often extremely difficult, and sometimes impossible, to reproduce exactly the conditions which would be met in actual service. For example, the neutron flux in a power reactor might be substantially greater than can be found in even the most powerful research reactors. The chemist must therefore be prepared to carry out experiments of several months' or years' duration, and to devise methods of measuring the small changes which occur during the early stages of the reaction.

(iii) Processing of Spent Fuel Elements

After a certain period of irradiation fuel elements are removed from the reactors and processed in order to remove plutonium and the fission products. Obviously, this aspect of reactor technology lies very much within the domain of the chemist. A solvent extraction process was selected for the Windscale factory, but before the design could be started the chemistry of plutonium—an entirely new element—had to be investigated. Research is now being directed towards improving this process and developing radically different processes which promise to have several advantages over solvent extraction.

(iv) Disposal and Utilisation of Fission-product Wastes

The fission products, which are removed in the processing of spent fuel elements, will be produced in increasingly large amounts as time goes by. It is necessary that these be disposed of safely, or, preferably, put to some useful purpose. Several methods of achieving these objects are being actively investigated.

(v) Fundamental Work

The Atomic Energy Authority encourages research work of a fundamental nature, much of which is centred at Harwell. For example, the fundamental chemistry of many of the processes used in the factories, such as solvent extraction, has been investigated to provide adequate information for the satisfactory design and operation of the plants. Attention is not only confined to topics closely related to the factory processes. At Harwell unique facilities are available; these include such expensive items as nuclear reactors, particle accelerators and buildings designed for the handling of toxic radioactive materials. Chemists make wide use of many of these facilities in studying the fundamental chemistry of the process of fission and of the transuranic elements, and in studying the chemical effects of high energy radiations.

Developments in each of these fields will be elaborated in later chapters of this book.

THERMONUCLEAR REACTORS

At sufficiently high energies light nuclei, such as H, D or Li, fuse with the release of a large amount of energy. One means of achieving the necessary particle energies is to heat the light atoms to extremely high temperatures, of the order of millions of degrees. Such a fusion process is achieved in the thermonuclear bomb, where a fission bomb is used to provide the necessary temperature, and it is believed to be the source of energy in the sun and stars. Temperatures of 5 million degrees have been achieved in the experimental thermonuclear apparatus ZETA at Harwell. Here deuterium gas at a low pressure is heated by passing through it a high electric current. Although fusion has not been observed at these temperatures, it is anticipated that considerably higher temperatures will be obtained after suitable modifications to the apparatus. At present the high temperatures can be sustained for only a few milliseconds and the energy released is insignificant compared with the energy input required.

Extensive development, possibly over several decades, will be necessary before a power-producing reactor can be considered. Even in such an event, the economics of the system are at present completely unknown. Although the fuel, deuterium, is abundant and can be extracted fairly cheaply from water, the capital costs of a thermonuclear reactor are likely to be high. The time-scale involved in developing an economic system, even if this does become a practical reality, is such that nuclear reactors based on the fission of U^{235} and Pu^{239} will remain an important feature of the industrial scene for many years to come.

It is not possible to define, at such an early stage, the part the chemist will play in the development of a thermonuclear fusion reactor. Very likely his contribution will be limited to the production of pure fuel materials and possibly the development of suitable materials resistant to high temperatures, for the walls of the system. In the process of fusion itself chemistry has no meaning, for at the temperatures involved (up to $10^{8}°C$) molecules do not exist (the product RT is of the order of 2.10^{5} kcal, compared with chemical bond-energies of 10 to 100 kcal). Even at the temperatures attained so far ($5.10^{6}°C$) hydrogen and deuterium are fully ionised, and oxygen and nitrogen are ionised to the $+4$ and $+5$ states respectively. Moreover, unlike the fission products, the products of fusion (e.g. He^4) are not radioactive and therefore will not necessitate elaborate chemical separation and disposal processes.

FURTHER READING

Jay. *Britain's Atomic Factories.* Her Majesty's Stationery Office, 1954.
Jay. *Atomic Energy Research at Harwell.* Butterworths Scientific Publications, 1955.
U.K.A.E.A. *Annual Reports.* Her Majesty's Stationery Office, 1955 onwards.

CHAPTER 2

RAW MATERIALS

EVER since the Industrial Revolution, Great Britain, followed by many other countries, has been squandering her energy reserves in the form of fossil fuels at an ever increasing rate in a multitude of thermodynamically inefficient processes. Technological advances have been able to increase neither the supply nor the thermodynamic efficiencies to the extent necessary to keep pace with increasing demands, and it has been possible to predict the complete exhaustion of fossil fuels within the foreseeable future.

The atomic energy industry was born at a time which will allow it to take over the main energy production within the next two decades in those countries which have low reserves of economically recoverable coal and which have an extensive grid system for electricity distribution. The remaining coal can then be used as a source of valuable chemicals rather than as a fuel. The extension of fuel resources which this implies will be somewhat different for each country, according to whether she happens to be the fortunate possessor of extensive deposits of natural uranium or thorium, or whether she has favourable trade relations with other uranium-producing countries. In the most favourable cases the resources of uranium and thorium should be sufficient to supply all energy needs for the next few hundred years. Long before that time, of course, it is expected that fusion power based upon the essentially unlimited supply of deuterium in the oceans will become available.

It is the purpose of this chapter to describe briefly how the natural resources of uranium and thorium are distributed and what methods are used for their recovery.

With the exception of a very few experimental fast reactors, all the fission reactors built to date have required a moderator to slow down the neutrons to energies at which they can be efficiently captured by U^{235}. The most widely used moderators are heavy water and graphite. Many reactors are based upon ordinary water for neutron moderation but considerable nuclear advantage can be gained from the use of separated deuterium (heavy water). The wastage of neutrons by absorption is then less than in light water, and the U^{235} content of the fuel is correspondingly smaller. Chemical aspects of the production of heavy water and of graphite are included in this chapter.

URANIUM

OCCURRENCE

Uranium is one of the less common elements to be found in the earth's crust. Estimates of its abundance have varied somewhat, but they appear to be of the order of a few parts per million. This is about equal to the abundance of tungsten and tantalum and is somewhat less than that of hafnium, arsenic and beryllium.

Minute traces of uranium are widespread. The oceans, for instance, have been found to have a uranium concentration of 3 μg/l. measured at the surface. Such a dilute source of fissile material has not been exploited up to the present time since sufficient uranium has been obtained from the more concentrated sources which nature supplies.

Uranium is also fairly widely distributed at very low concentration in some of the more common rocks. Granite, for instance, contains a small quantity of uranium, but it would be necessary to process over 100 tons of rock to obtain one pound of uranium oxide. Phosphate rock would be somewhat better, having a uranium oxide content of one pound to about 4 tons of rock.

The uranium content above which a uranium-bearing ore becomes commercially useful depends upon the ease with which the ore can be mined (very extensive low grade deposits near the surface will compete with higher grade material at a greater depth), upon the development of efficient extraction processes, and upon the price which the constructor of atomic power stations is prepared to pay.

Each one of the early gas-cooled reactors in the United Kingdom nuclear power programme will require an initial uranium investment of several hundred tons. This fuel charge will cost up to £5 million and will be a large fraction (20 to 25 per cent) of the total cost of the complete power unit. Moreover, in the power stations which are due for completion in the early 1960's, the cost of replacing the burnt fuel will amount to approximately one-third of the cost of each unit of electricity sent out. These high fuel-element costs are divided into two major items:

Cost of ore concentrate	High, about £10 000/tonne uranium
Extraction and purification	Low
Manufacture of fuel elements	High, approaching £10 000/tonne

The discovery of uranium deposits which are cheaper to mine and to concentrate will obviously have a significant effect upon the cost of power produced.

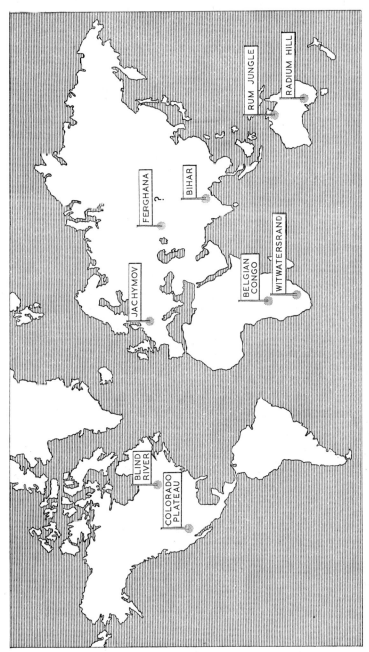

FIG. 3. PRINCIPAL URANIUM-PRODUCING AREAS

To match the figure of £10 000/tonne of uranium in the ore concentrate it is generally considered profitable to work ores containing a lower limit of 0·1 per cent U_3O_8, or 2 pounds per ton of rock. The very extensive deposits at Blind River, in Canada, are only slightly above this limit of uranium content and the magnitude of the task involved in extracting the uranium is indicated by the fact that early in 1958 the rated capacity of the plants in this area was over 9 million tons of ore per year.

The South African ore grade is very low—only 0·01 to 0·03 per cent U_3O_8—but it has been found profitable to work this ore as a by-product of the gold industry.

The principal uranium-producing areas of the world are shown in Fig. 3. Australia has been included, since that country has announced a programme for future production which amounts to more than the Belgian Congo. India is intending to develop a new uranium field in the State of Bihar, with a known reserve of more than 15 000 tons of uranium at greater than 0·1 per cent concentration.

Uranium Minerals

By far the most important rock formations in which uranium minerals occur are those of the pre-Cambrian era. Within these rocks, the uranium can exist as a wide variety of chemical compounds, including oxides, silicates, phosphates, vanadates and sulphates. The oxide, uraninite, is the most widely reported of the uranium minerals. Pitchblende, which played such an important part in the early history of radioactivity, is a particular variety of uraninite with a very small particle size. Fairly large lumps of the mineral are often found; the record appears to be a single mass of oxide weighing 20 tons. The mineral contains 75 to 95 per cent uranium as the dioxide, UO_2.

The widespread, but very low grade, uranium deposits in the gold reefs of South Africa are composed primarily of uraninite and thucolite. It is interesting to note that although the uranium concentration is many times that of the gold content of the reefs, uranium production has remained secondary to the gold industry. It is primarily the gold content of the ore which decides whether or not the deposit will be worked and the uranium concentration plants have to take whatever ore passes through the gold plants. Despite this fact, the annual production of uranium at the end of 1956 was 5000 tons. The association of uranium with the gold-bearing areas appears to be somewhat coincidental, since there is no direct chemical or geological similarity between them.

There are two important multiple-oxide minerals of uranium, davidite and brannerite. The former mineral is a mixture of the oxides of iron, titanium and the lanthanides, containing about 10 per cent of uranium. It is less important than brannerite which contains about 40 per cent uranium oxide associated with titanium dioxide and traces of thorium, the lanthanides, calcium and iron. Brannerite, together with uraninite, forms the basis of the large Blind River deposits in Canada, whereas davidite occurs at Radium Hill in South Australia.

These primary ores remain stable in chemically reducing environments, particularly if they remain below the level of the water table. More oxidising conditions normally exist above the water table and this has led, for instance, to the conversion of black uraninite into the secondary minerals, green torbernite ($CuO.2UO_3.P_2O_5.8H_2O$) and yellow autunite ($CaO.2UO_3.P_2O_5.8H_2O$). Similarly, uranium originally present as uraninite associated with vanadium minerals in the Colorado Plateau region has been transformed by oxidising conditions to secondary carnotite ($K_2O.2UO_3.V_2O_5.3H_2O$).

Uranium is often associated in small concentrations with carbonaceous materials. Thucolite is a mineral which contains uranium, thorium, carbon and water. The manner in which the mineral was originally formed is still obscure. It occurs with the other uranium minerals, for instance, at Blind River and in the Witwatersrand deposits, and in isolation in bituminous shales. The shale deposits in Sweden are estimated to provide a uranium reserve sufficient to maintain a substantial atomic energy industry in that country, provided that suitable methods can be evolved for recovery from this rather low grade ore.

RECOVERY

Unlike many of the more common metals, such as iron, uranium is not extracted by smelting the ore, but by a process of dissolution and purification.

There are three major stages in the recovery:

Pre-concentration
Leaching
Separation

The details of the pre-concentration stage differ according to the particular mineral which is being worked, and upon the state of subdivision of the mineral. Physically, the pre-concentration step involves crushing, grinding, and the separation of much of the unwanted gangue. Before commencing the leaching stage it is necessary also to ensure that the uranium is in the water-soluble hexavalent state.

Depending upon local circumstances, the uranium oxidation is carried out by the addition of manganese dioxide, ferric ion or chlorate ion, to the aqueous slurry of ore concentrate. In the recovery of uranium from South African gold-plant residues, pulverised naturally-occurring manganese dioxide is added. Ferric iron is believed to be necessary for the oxidation: sufficient iron enters the treatment vessels by the process of abrasion of steel balls and cylindrical mill linings in the gold treatment.

There are two main leaching methods for removing the oxidised uranium from the ore concentrate: use of sulphuric acid or use of alkali carbonate solutions. Sulphuric acid leaching produces a solution of uranyl sulphate, whereas the carbonate leach produces the soluble complex ion $UO_2(CO_3)_3^{4-}$. The choice between these methods is dependent upon mineral content of the ore.

The carbonate leach is more selective than the sulphuric acid process, owing to the fact that most metal carbonates are insoluble in water and do not form soluble complexes. Either high temperatures or high pressures are needed for the carbonate leach, however, in order to speed up the solution process. Moreover, the process has a large requirement for carbonate and is successfully applied only to ore concentrates which contain substantial amounts of alkaline-earth carbonates. If the sulphur content of the concentrate exceeds 0·5 per cent (for instance, due to the presence of pyrites) then it is necessary to remove this at the pre-concentration stage since the sulphide minerals, which are decomposed by the hot alkaline solutions, would become the principal consumers of reagents. It has been claimed that ammonium carbonate has some advantages in avoiding the side-reactions of added sodium carbonate, but its use requires high-pressure leaching techniques to avoid decomposition of the product to an insoluble oxide by the reaction:

$$(NH_4)_4.UO_2(CO_3)_3 \rightarrow 4NH_3 + 3CO_2 + UO_3.2H_2O$$

When the ore concentrate contains silicious material and only a low proportion of natural carbonates, sulphuric acid leaching is generally preferred for economic reasons. One of the disadvantages of the sulphate process is that the plant must be built of acid-resisting materials and this significantly increases the cost.

Whereas the presence of sulphides in the ore was an embarrassment for the alkaline leach process, it can be turned to good account for acid leaching. Recent work has shown that the oxidation of sulphides in aqueous slurries at 130°C by compressed air proceeds rapidly, and the sulphuric acid which is produced may be used for the dissolution of the uranium mineral *in situ*. This may become an important method for uranium recovery in the future, since the only reagent required would be compressed air, plus perhaps some excess sulphide ore if the sulphide content of the uranium ore is too low.

When the uranium has been converted into a soluble form by one of the two primary leaching procedures, there remains a choice of two distinct types of further process. In one process the uranium liquor is separated from the spent residue by filtration, and the uranium is subsequently recovered by precipitation, adsorption on anion-exchange resins, or solvent extraction. The alternative route, which may find wider application in the future, involves extracting the uranium without prior separation from the spent residue. Two methods of achieving this have been under investigation: contact between an ion-exchange resin and the aqueous ore-pulp in a counter-current extraction vessel, and direct solvent extraction of the ore-pulp.

The precipitation process for recovering uranium from the separated liquor, or from the back-wash solutions of direct extraction methods, is complicated by the co-precipitation of iron. Normally the precipitation of ammonium diuranate, $(NH_4)_2U_2O_7$, is achieved by the addition of ammonia, which also precipitates ferric hydroxide. In the South African plants, however, separation of uranium from iron in the liquor from the sulphuric acid dissolution is achieved by first precipitating the iron with the addition of milk-of-lime until the liquor pH reaches 3; after filtration, the uranium is precipitated with ammonia at pH 6·9.

At this stage, the product has a composition approaching that of a chemical compound of uranium. The impurity levels are still far too high for nuclear use, however, and further purification is required, usually after the material has been shipped away from the producing areas to the consumers.

Springfields Factory

A rather different procedure was adopted for the first plant erected in 1947 at the Springfields factory in the United Kingdom. Here the crude uraninite ore was received direct from the Belgian Congo and the extraction and purification was performed in the same factory as the final metal production. See Plate I.

An acid leach was used and a further refinement was added at this stage in order to separate the radium which occurs with the

uranium. Radium is a daughter element formed during the
radioactive decay of uranium:

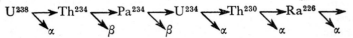

$$U^{238} \xrightarrow{\alpha} Th^{234} \xrightarrow{\beta} Pa^{234} \xrightarrow{\beta} U^{234} \xrightarrow{\alpha} Th^{230} \xrightarrow{\alpha} Ra^{226} \xrightarrow{\alpha}$$

It occurs, therefore, in all uranium ores, the equilibrium concentration being about one part in three million. Most of the radioactivity
associated with the original ore is due to these daughter products
rather than to the parent uranium. It was the observation that
the activity could be separated chemically from the uranium which
lead to the first isolation of radium from pitchblende by M. and
Mme Curie at the turn of the century. The radium was separated
at Springfields by co-precipitation with barium carbonate: the
uranium remained in solution as a complex ion.

The radium content of the ore, prior to this separation, was the
main factor in determining the health precautions which must be
taken. The daughter product of the alpha particle emission from
radium is radon, a gaseous element. During the process of crushing
the ore, radon which has been trapped within the mineral crystals
is released and must be prevented from entering the atmosphere
within the process building.

One difference from the South African practice was the use of
nitric acid addition to the sulphuric acid at the leaching stage as an
oxidant for the uranium. This had the additional advantage of
providing a corrosion-resistant oxide film on the steel equipment.

After the acid leach, uranium was precipitated from the filtered
aqueous solution by the addition, not of ammonia as in the South
African industry, but of hydrogen peroxide. Some ammonia was
necessary to control the pH of the solution, but the precipitate was
a hydrate of UO_4 rather than the ammonium diuranate which
would result from the use of ammonia alone. A better decontamination from impurities was thereby achieved.

Hydrogen peroxide is a potentially hazardous chemical to
handle and its use had not been attempted on such a scale in the
United Kingdom prior to the operation of the uranium production
plant. Although it gave very satisfactory decontamination of the
uranium from impurities, its introduction gave rise to complications in the operation of the plant. It was found necessary, for
instance, to exercise careful control of the solution temperature at a
level somewhat below that of the ambient air in order to reduce
the rate of decomposition of hydrogen peroxide catalysed by the
presence of iron impurity in the uranium solution.

More recently the feed for the Springfields plant has been in the form of uranium concentrates from South Africa, Australia and the Belgian Congo, rather than the original uraninite ore. These concentrates are of much lower impurity levels and it has been found possible to replace the peroxide precipitation step by the economically more attractive precipitation with ammonia.

The uranium peroxide precipitate, or the ore concentrate, is dissolved in nitric acid and then undergoes further purification to the high standards required for reactor use by means of solvent extraction. Uranyl nitrate is soluble in diethyl ether to a much greater extent than are the impurities, consequently contact between the aqueous nitrate solution and ether preferentially extracts uranium. After separation of the ether solution, the uranium is returned to aqueous solution by contact with a large quantity of demineralised water and at this stage the uranium is in its most pure state. The further stages in the production of metallic fuel elements all introduce new impurities in small amounts, despite elaborate precautions.

A recent innovation, in the direction of higher extraction efficiency and lower potential operating hazard, has been the replacement of ether as the organic extractant by tributyl phosphate in an inert hydrocarbon diluent. A somewhat similar situation also exists in the United States, where the early plant installed at the Mallinckrodt works still uses ether extraction, whereas the more recent National Lead Company plant at Fernald employs tributyl phosphate.

The purified uranium is precipitated from solution by adding ammonia to form ammonium diuranate. This is later dried at about 350°C and roasted under reducing conditions (hydrogen at 700°C) to give the dioxide. The latter is converted to the tetrafluoride with gaseous hydrogen fluoride at elevated temperatures and the final reduction to metal is achieved by reaction with a metal such as calcium:

$$(NH_4)_2 . U_2O_7 \xrightarrow[350°C]{H_2} UO_3 \xrightarrow[700°C]{} UO_2 \xrightarrow{HF} UF_4 \xrightarrow{Ca} U$$

The conversion from ammonium diuranate to uranium tetrafluoride has been performed as a batch operation up to the present time, the uranium compounds being held in trays and the gases being passed over them. Future developments may introduce continuous operation by means of fluidisation techniques. Some further discussion of the reduction of uranium tetrafluoride to the metal appears in Chapter 4.

C.N.P.   c

Not all of the Springfields uranium tetrafluoride is converted directly to metal: some of it is required to provide the uranium hexafluoride feed material for the Capenhurst diffusion plant. The latter employs the principle of gaseous diffusion through porous membranes to separate U^{235} from natural uranium. The separated isotope can be used for military purposes or to enrich the fuels for power reactors.

The need for enrichment of fuel most commonly arises from the necessity to employ reactor core constructional materials of relatively high neutron absorption, or from the desire to make the smallest possible reactor system for propulsion purposes. Very little detailed information has been released concerning the operation of the diffusion plant itself, although methods for converting uranium tetrafluoride have been widely reported in the literature.

The Springfields process employs liquid chlorine trifluoride for the conversion:

$$3UF_4 + 2ClF_3 \rightarrow 3UF_6 + Cl_2$$

The reaction may be more readily controlled by using the liquid fluorinating agent rather than gaseous fluorine or bromine trifluoride.

THORIUM

If we consider only the fissile (U^{235}) content of natural uranium, it is possible to equate the burning of one ton of uranium in nuclear reactors to the energy equivalent of about 21 000 tons of coal. Plutonium is produced from the U^{238} at the same time, and this gives us a credit factor which makes the implied fuel reserves even more promising. The world reserves of uranium are by no means infinite, however, particularly of the grades which are economical to work.

World energy reserves can be extended very considerably by the use of thorium. Natural thorium contains primarily the isotope Th^{232}. This does not undergo fission with thermal neutrons, but it has an appreciable neutron absorption cross section to give Th^{233}, which then decays by beta particle emission through protactinium to U^{233}. From a nuclear physics point-of-view, the latter isotope is the most desirable for use in thermal reactor systems. It may be possible to build *breeder* reactors, for instance of the homogeneous aqueous type, which operate with U^{233} fuel in the core and have a surrounding blanket of Th^{232}. If the technological problems can be overcome these reactors could produce up to 1·1 atoms of U^{233} in the blanket for each U^{233} atom consumed in the core, giving a gain of 10 per cent of fissile material which could be used to fuel further reactors. Thorium would become the feed material for such a reactor cycle.

OCCURRENCE

Thorium is widely distributed in the earth's crust; at an average concentration of about 12 parts per million it is some three times more abundant than uranium. It occurs as a trace constituent in many minerals, but only six minerals are known in which it is an essential component. This is far less than the number of uranium minerals and reflects the much simpler chemistry of thorium.

The primary minerals of thorium, being stable against oxidation and very insoluble in water, show geological features quite distinct from those of uranium. The largest known reserves of thorium are detrital deposits formed by erosion of older rocks and transportation and deposition of the insoluble mineral particles.

Thorium Minerals

Thorianite (ThO_2) is not as widely distributed as uraninite (UO_2) and is not a promising large-scale source of thorium. Thorite is somewhat more important; it has the ideal formula $ThSiO_4$, but this is always considerably modified by the presence of a number of other elements—uranium, iron, calcium, the lanthanides and phosphates, in solid solution.

These minerals are completely overshadowed by monazite as a potential source of thorium. Monazite sand is a complex mixture of the phosphates of cerium, yttrium, lanthanum and thorium; the thorium content is highly variable. Being chemically stable, insoluble and hard, the monazite has been eroded from igneous rocks, and transported by streams to form alluvial deposits. If these deposits are at a sea coast, further movement and concentration occurs by along-shore currents.

The beach sands of the Indian Malabar and Coromandel coasts are very rich in monazite and are estimated to contain up to 180 000 tons of thorium in an easily extractable form. An important thorium reserve further inland has been discovered also, in the State of Bihar; this is estimated to contain over 300 000 tons of thorium at a concentration of over 10 per cent.

Other large scale deposits of monazite occur in Brazil. Much more diluted, and smaller, deposits have been reported in western Taiwan, Ceylon and the U.S.A. The distribution of thorium in the U.S.S.R. has not been reported.

RECOVERY

Monazite has been used as the source of thorium up to the present time. It may contain up to 10 per cent of thorium, associated with about 60 per cent of the rare earths and a fraction of a per cent of uranium. The deposits are often coloured black by the presence of magnetite, or red by the presence of garnet.

We have said above that monazite is characterised by its chemical stability, and in order to recover the thorium it is necessary to break down the structure of the sand by digestion with hot concentrated sulphuric acid. To obtain the thorium in solution it is necessary to employ considerably more acid for this process than would be required for stoichiometric chemical reaction, owing to the rather low solubility of thorium sulphate.

Separation of the thorium from the solution containing sulphate, phosphate and the rare earths is not easy, but a variety of methods is available. Early separations were made by fractional precipitation or crystallisation, but these tedious processes are likely to be superseded in future operations. Carbonate precipitation has received most attention; addition of an excess of sodium carbonate to the crude thorium solution precipitates the rare earth carbonates, but the thorium remains in solution as a complex ion.

A solvent extraction process may well be developed for thorium separation, tributyl phosphate being a particularly promising solvent. This extracts the thorium and leaves the rare earths in the aqueous phase; the presence of sulphate ions in the solution reduces the extraction coefficient, but even here a satisfactory separation has been shown to be possible.

Thorium metal is produced from the purified thorium stream by reduction of the oxide with calcium or the fluoride with a mixture of zinc chloride and calcium (see Chapter 4).

GRAPHITE

The fission cross-section of natural uranium increases from about 0·015 barn at a neutron energy of 1 MeV to 3·9 barn at thermal neutron energies. To produce a neutron chain reaction in natural uranium it is necessary to add a moderating material which will slow down the neutrons by collision with the moderator nuclei. The principal moderators which have been used or proposed are water, D_2O, beryllium, BeO, and graphite.

Graphite was used in the first chain-reacting pile in Chicago and has been the basis of many reactors since that date. It is relatively cheap, easy to machine, and has a lower neutron capture than water.

Natural deposits of graphite are known, particularly in Ceylon,

U.S.A. and Madagascar. The natural material has not been used for reactor construction, however; one of its disadvantages is a poor mechanical strength compared with artificial graphite.

Depending upon the required purity and physical properties, artificial graphite can be prepared from a wide variety of organic materials. However, reactor-grade graphite is usually manufactured from petroleum coke as a raw material; special attention is given to sources of the coke which have a low impurity content, particularly boron.

Petroleum coke is produced at oil refineries as a residue from the distillation of heavy oil. It is roasted to about 1300°C in order to remove all but a small fraction of volatile hydrocarbons. The calcined coke is then crushed and blended with about 30 per cent of pitch residue from coal tar distillation. The pitch acts as a binder during the subsequent processing. The required shapes of graphite are formed by either extruding or by moulding the blended material.

These two methods of forming give rise to graphite with quite different physical properties. The crushed coke particles are long and relatively thin owing to the fact that in the coking process the aromatic molecules mainly become oriented with the planes of the benzene rings parallel to the cellular walls of the coke. The normal characteristics of fracture of these walls during crushing is such that the length of the particles is in the direction of the layer planes of the ultimate graphite structure. During extrusion many of the coke particles become aligned parallel to the direction of extrusion. This donates a grained structure to the final graphite, with the grains lying parallel to the direction of extrusion. The graphite will thus have higher thermal and electrical conductivities along the direction of extrusion than across it.

In moulding, however, the coke particles, being long and thin, tend to become forced by a purely mechanical process into a position with their longest dimension normal to the applied force. The thermal and electrical conductivities of the final graphite will then be higher in the directions normal to the moulding force. Consequently the extrusion process gives approximately equal properties in two directions mutually at right angles, but with higher strength, conductivity and thermal expansion along the third axis; whereas moulding produces graphite with lower values of these properties along the third axis.

After the forming process the coke-pitch blocks are heated in a gas-fired furnace to about 750°C and the pitch binder evolves large quantities of hydrocarbon gases. The blocks undergo considerable shrinkage between 450 to 750°C.

The reactor designer requires a graphite with a high density. In order to attain this, the fired blocks are further impregnated with pitch. One means of carrying this out is to heat the blocks to 250°C in an evacuated vessel and then to pump pitch at a pressure of 100 p.s.i. into the vessel.

The final graphitisation process is then performed by prolonged heating of the blocks to 2600° to 3000°C in a furnace using a coke packing as the resistance heating element. It is fortunate that many of the impurities vaporise away from the graphite during this treatment.

Purity requirements for the graphite for reactor use are so stringent that chemical analysis of the product does not constitute a sufficiently sensitive detection method. Materials such as the rare earths with very high neutron-absorption cross-sections can have a significant effect upon the apparent graphite cross-section at extremely low concentrations. To detect these it is necessary to determine the effect of graphite samples upon the nuclear reactivity of a test reactor. The first reactor to be built in the United Kingdom—GLEEP—has been used for this purpose.

Natural graphite has a density of 2.26 g/cm^3. Artificial graphite has a lower density (1.6–1.7 g/cm^3); the particular value depends upon the starting materials and upon the details of the manufacturing process. This lower density implies a comparatively high porosity, which in turn is a considerable disadvantage since it increases the amount of surface available to take part in oxidation reactions. One of the primary aims for the future will be the production of graphite with a much lower porosity, to minimise the reaction with reactor coolant gases and to prevent the passage of fission product gases from the fuel to the coolant in the high-temperature gas-cooled reactor (Chapter 8). Success in this aim will probably depend upon alteration of the surface of formed graphite material; manufacture of large blocks with a low porosity uniformly distributed is difficult, because this low porosity prevents the escape of hydrocarbon gases during the heating cycles used for manufacture.

DEUTERIUM AND D$_2$O

Heavy water has a good moderating power for fast neutrons, combined with a low neutron-absorption cross-section. Moreover, since it is a liquid, it does not suffer radiation damage to the same extent as graphite; for example, it does not give rise to the Wigner energy storage which occurs in graphite. It is, however, considerably more expensive to manufacture than is graphite and one of the primary aims in the design of heavy water moderated reactors is to cut down the heavy water investment to the absolute minimum. Heavy water is used in research reactors such as NRX at Chalk River and DIDO and PLUTO at Harwell. It is proposed as the working fluid in most forms of the homogeneous aqueous reactor (Chapter 7) and it is essential to the heavy-water moderated gas-cooled reactors which are under consideration in Sweden, Canada and the United Kingdom. A single, large, heavy-water moderated, gas-cooled reactor (1000 MW, heat) would require about 250 tons of D_2O.

Deuterium may find an important use as an energy source also in the future. Nuclear fusion reactions occur at a much lower temperature in deuterium than in hydrogen, and the problem of sustaining a fusion reaction in deuterium is thereby rendered somewhat easier of solution. Although early research on the fusion process, in apparatus such as ZETA, has shown great promise, there are many practical difficulties remaining to be solved before it can be stated that power from fusion is likely to be cheaper than power from coal or from fission. Nevertheless we may look forward confidently to the time when these difficulties will have been overcome and deuterium becomes our major reserve of power. Deuterium occurs in the water of the oceans at a concentration of not more than 150 parts per million, but the extent of the oceans implies an essentially inexhaustible supply. The development of new, cheaper, methods for the recovery of deuterium may be one of the chemist's and chemical engineer's most important contributions to the success of the fusion-reactor programme.

There are many ways in which the differences in the properties of hydrogen and deuterium, or their compounds, are sufficient to merit consideration as a separation process. The methods which may be used economically for large scale production are severely limited, however, and may be listed as three types: water electrolysis, distillation and chemical exchange.

Small amounts of heavy water (a few tons per year) are produced most economically as a by-product; the major product bears the brunt of the costs of erecting and maintaining the plants. An upper limit is set by the desired maximum output of the major product, however. As an example, the plant at Rjukan, Norway, which began operation before the war, produced 1·7 tons of D_2O per year as a by-product of hydrogen generation for the synthesis of 100 000 tons per year of ammonia. Similarly, all new large fertiliser plants to be built in India will be designed to produce heavy water also: one plant coming into operation in 1960 will give

340 000 tons per year of nitrogenous fertiliser plus 10 to 20 tons per year of D_2O.

If large nuclear-power reactor systems are to be based on heavy water as a moderator, and if substantial quantities are required for fusion-power production, then the demand will exceed the output of these by-product plants and it will be necessary to design plants specifically for the purpose of deuterium production. Some such plants have been operated in the United States in order to fulfil a military requirement.

SEPARATION PROCESSES
Water Electrolysis

In thinking of the electrolytic dissociation of water into hydrogen and oxygen one does not normally consider the fate of the very small concentration of deuterium which is always present. It has been found, however, that the deuterium content of the hydrogen evolved at the cathode is somewhat lower than that of the water remaining in the electrolytic cell. As electrolysis continues, the residual water becomes progressively enriched in deuterium.

A comparatively high separation factor can be attained by this method, but, because of the low initial deuterium content, it is necessary to process enormous quantities of water to reach an output of more than a few tons of deuterium per year. Moreover the process can only be contemplated when large quantities of cheap (hydro-electric) power are available.

The first commercial production of heavy water was by the electrolytic process used at Rjukan by the Norsk Hydro Company. This plant used many electrolysis cells in parallel-connected stages which were connected in series cascade. The water fed to each stage was about 73 per cent electrolysed and the remaining 27 per cent was carried out as entrained water vapour in the hydrogen and oxygen stream. The water vapour was condensed and fed to the succeeding stage for further electrolysis. By this means the deuterium content of the water was concentrated from about 130 parts per million to 15 atom per cent. Final purification was achieved in a secondary plant which operated on the same basic principle. The electrolytic method has been used in some installations, for instance at Trail, British Columbia, as a means of refining crude heavy water.

Distillation

Several distillation processes are possible: the primary requirement is that the vapour pressure of some compound of deuterium should differ appreciably from that of the corresponding hydrogen compound. Such differences are quite small—the smaller the difference, the less likely is the process to be industrially attractive. As a specific example we may quote the relative volatility for isotope separation from water at 100°C as 1·026.

Water Distillation. The separation factor achieved during distillation of water is not a straightforward ratio between the vapour pressures of D_2O and H_2O. The situation is complicated by the presence of the species HDO. However, separation can be achieved and most of the early D_2O produced in the U.S.A. was made by this method. Just as in the electrolytic method it is necessary to supply large quantities of energy to the process in the form of electricity, so in the water distillation method large quantities of heat must be supplied. It has been estimated that one mole of D_2O requires 200 000 moles of steam. Water distillation plants were operated in the early days of the United States Manhattan District project, but were later closed down on economic grounds.

Hydrogen Distillation. Hydrogen contains the species H_2, HD and D_2. The ratio of the volatilities of liquid H_2 and HD is higher than that for any other distillation process, being 3·6 at the triple point temperature of − 259·4°C. It is the novel engineering features of the design of large plants to work at these extremely low temperatures which have so far prevented the manufacture of deuterium by this method. However, a plant is now being built in France to work on this principle and further plants are contemplated in India.

It is not economical to produce hydrogen solely for the separation of deuterium. Very large quantities of hydrogen are used in industry, for instance for synthetic ammonia production, and it is envisaged that deuterium could be separated from the main hydrogen stream in certain circumstances—high annual output and low fuel costs.

If the hydrogen for a synthetic ammonia plant is produced as water gas rather than by water electrolysis, then elaborate precautions would be required to remove impurities before attempting to separate the deuterium. The complexity of the purification has been discussed by Manson Benedict. Most of the carbon dioxide content would be removed as a first step by scrubbing the gas mixture with caustic soda. Oxygen may be removed by recombination with the hydrogen when the gas stream is passed over a catalyst bed—for example, palladised alumina. Nitrogen, water vapour and more carbon dioxide would be removed next by cold traps, followed by adsorption of methane and the last traces of carbon dioxide on silica gel. Last traces of nitrogen would then be taken out as the solid by cooling to — 232°C.

Deuterium is removed from the purified hydrogen by fractional distillation at — 250°C. The first distillation column produces hydrogen almost free of deuterium at the top and a concentration of 5 to 10 per cent HD at the bottom. The latter is then fed to a second column for further purification. Finally the HD is disproportionated into H_2 and D_2 over a catalyst, and this mixture is separated in a further distillation. The hydrogen and nitrogen, after being used to cool incoming gases in heat exchangers in earlier parts of the process, are both returned to the synthetic ammonia plant.

An interesting problem arises from the possible conversion of hydrogen from the ortho-form to the para-form. The hydrogen feed to the separation plant would contain approximately 75 per cent of ortho-hydrogen. The conversion to para-hydrogen is an exothermic process; if the conversion were allowed to go to completion the amount of heat liberated would exceed the refrigeration capacity of the plant. The silica gel which has to be used to remove last traces of the impurity gases has been found to show considerable catalytic activity for this conversion, particularly if it contains adsorbed oxygen. Consequently the plant must be designed with the minimum possible amount of silica gel, and the oxygen removal step in the hydrogen purification must be exceedingly efficient.

Chemical Exchange

Three isotopic exchange reactions have been considered for deuterium concentration. They are the equilibria:

$$H_2O + HD \rightleftharpoons HDO + H_2 \qquad (a)$$
$$H_2O + HDS \rightleftharpoons HDO + H_2S \qquad (b)$$
$$NH_3 + HD \rightleftharpoons NH_2D + H_2 \qquad (c)$$

Attempts are being made to design dual-temperature closed-cycle plants so as to free the production capacity from dependence on hydrogen (or H_2S) produced for other purposes. Thus, in the steam/hydrogen exchange, the hydrogen would be recycled and the raw material fed to the plant would be steam.

Steam-hydrogen Exchange. Equilibrium constants for reaction (a) are 2·55 at 100°C and 1·30 at 600°. The closed-cycle dual temperature system takes advantage of this difference in the equilibrium. In a reaction vessel kept at the lower temperature there is a tendency for transfer of deuterium from the hydrogen to the steam; the steam is condensed and leaves the cycle, so concentrating the deuterium. The hydrogen is then fed to a reaction vessel at the upper temperature. Here the deuterium-depleted hydrogen picks up more deuterium from another feed of *natural* steam and the hydrogen is recycled to the low temperature reaction vessel. Although steam-hydrogen exchange has been used to concentrate the deuterium-rich product from water electrolysis, large scale application has not been shown to be economically feasible. Primarily this is due to restrictions imposed by the necessity for a suitable catalyst. Several compounds have been shown to have catalytic activity, for instance zinc chromite or nickel tungstate, but all of them are poisoned very easily by the presence of liquid water. The development of a catalyst which would remain active in the presence of liquid water would render the design of the plant much simpler and lead to lower estimated costs of production.

Water-hydrogen Sulphide Exchange. Contact between liquid water and H_2S gas in a counter-current reaction tower at 25 to 30°C leads to the water being enriched at the expense of the H_2S. The separation factor is temperature dependent. Consequently a portion of the enriched water may then be transferred to a higher temperature tower where its deuterium content is transferred to the H_2S. The plant has, therefore, a feed of cold water which is converted to a deuterium-rich product stream, and there is a waste stream of deuterium-depleted water. The H_2S is recycled and it is only necessary to add small quantities to make up for losses.

In order to obtain a high separation factor the plant would be designed with a hot tower temperature up to 130°C, requiring some pressurisation to maintain the water in the liquid state. This process has an advantage over the steam-hydrogen exchange in the fact that no catalyst is required. The major obstacle to large scale production of cheap heavy water by this method is the necessity to employ high-grade constructional materials to overcome the corrosive nature of the liquids.

Ammonia-hydrogen Exchange. Reaction (c) does not have the corrosion difficulties of the H_2S-water exchange, but it does require a catalyst to promote the attainment of equilibrium. No really suitable catalyst appears to have been found although potassium amide dissolved in the liquid ammonia shows some promise. Potentially this system has a higher separation factor than either of the other two processes.

The ammonia-hydrogen exchange could be used on *open-cycle* operation for stripping deuterium from large sources of industrial hydrogen. To free deuterium production from dependence on the hydrogen output it is possible to devise a closed-cycle dual-temperature exchange, but it is then necessary to couple this with a steam-hydrogen exchange tower at 600°C to transfer deuterium from water to stripped hydrogen.

An attempt has been made in this section to describe the basic features of several processes for deuterium separation. The decision as to which process is used may well differ according to local circumstances—for instance the cost and scale of hydrogen production. Each process has chemical problems which remain to be solved in order to effect a substantial gain in economy of operation.

FURTHER READING

Grainger. *Uranium and Thorium*, George Newnes Ltd.   1958.
United Nations.   Proceedings of the First International Conference on the Peaceful Uses of Atomic Energy, Geneva, 1955.
  Vol. 6, *Geology of Uranium and Thorium.*
  Vol. 8, *Production Technology of the Materials used for Nuclear Energy.*

CHAPTER 3

FISSION AND THE FISSION PRODUCTS

INTERACTION OF NEUTRONS WITH MATTER

DURING the 1930's many experiments were carried out in which a large range of elements were bombarded by atomic particles with a view to causing penetration of the nucleus. The first successful artificial transmutation was observed by Cockcroft and Walton in 1932; on bombardment with protons of energy 0·125 MeV the nucleus of lithium was converted into two helium nuclei (or alpha particles). The quantities produced were minute, the transmutation being observed by the tracks produced by the individual alpha particles in a cloud chamber. Positively charged particles suffer large coulombic repulsion forces on approaching an atomic nucleus, so that high particle energies, sufficient to overcome this repulsion, are required and even then the probability of success is very low. With neutrons, however, the reverse is usually the case since they have no charge. At high neutron energies (of the order of 1 MeV) the effective area of a nucleus is low, approximating to the geometric area of the nucleus, whereas, as Fermi discovered in 1934, at lower energies (of the order of 0·03 eV) the effective area rises to many times—a million times in some cases—the value at 1 MeV.

This effective area of the nucleus—or *cross-section*—is of paramount importance in reactor technology, for it is a direct measure of the degree of interaction between a flux of neutrons and the nucleus in question. The cross-section of a nucleus is usually expressed in units of the barn, one barn being 10^{-24} cm².

CONSEQUENCES OF NEUTRON CAPTURE

When an atomic nucleus captures a neutron, the atomic number (or the charge on the nucleus) remains unchanged but the atomic weight increases by one unit. Several alternatives are open to this compound nucleus, the relative importance of each alternative depending upon the nucleus in question and upon the energy of the incident neutron. These alternatives are:

(*a*) *Scattering*. The compound nucleus may re-emit the neutron in a random direction, so that neutrons from a neutron beam are scattered by the sample. In the scattering process energy is transferred to the bombarded nucleus to an extent which is inversely proportional to its mass, as in the collision of moving bodies obeying Newtonian mechanics. This process of degrading the neutron energy by collision with light nuclei—the *moderator*—will be seen to be of great importance in the operation of nuclear reactors.

(*b*) *Radiative Capture*. This is the most common process by which a change in the target nucleus takes place; the compound nucleus is usually in an excited state due to the change in the binding energy in forming the compound nucleus, and it decays to the ground state with the emission of one or more gamma quanta, the process being designated (n,γ) capture. The resulting nucleus may be unstable, decaying, with a characteristic half-life, by emission of a beta particle to give a new element. Some (n,γ) processes of importance in the field of atomic energy are:

$$H^1 + n \xrightarrow{(n,\gamma)} H^2 \text{ (deuterium)}$$

$$U^{238} + n \xrightarrow{(n,\gamma)} U^{239} \xrightarrow[t_{\frac{1}{2}}=23 \text{ min}]{\beta} Np^{239} \xrightarrow[t_{\frac{1}{2}}=2\cdot3 \text{ d}]{\beta} Pu^{239} \xrightarrow[t_{\frac{1}{2}}=24,000 \text{ y}]{\alpha} U^{235}$$

$$Th^{232} + n \xrightarrow{(n,\gamma)} Th^{233} \xrightarrow[t_{\frac{1}{2}}=23 \text{ min}]{\beta} Pa^{233} \xrightarrow[t_{\frac{1}{2}}=27\cdot4 \text{ d}]{\beta} U^{233} \xrightarrow[t_{\frac{1}{2}}=1\cdot6.10^5 \text{ y}]{\alpha} Th^{229}$$

The isotopes Pu^{239} and U^{233}, which do not occur to any extent in nature, are both of great value in the release of nuclear energy.

(*c*) *Emission of Charged Particles*. The emission of either a proton or an alpha particle is less common than the emission of a gamma photon, but does occur with some elements of low atomic weight, for example:

$$\text{(i) } B^{10} + n^1 \rightarrow Li^7 + \alpha^4 + 2\cdot5 \text{ MeV}$$

This reaction is of interest in providing a means of detecting and counting slow neutrons, as the $2\cdot5$ MeV of energy released in the process is carried by the charged Li^7 and alpha particle, which can be detected by the ionisation produced as the particles pass through a gas in an ionisation chamber.

$$\text{(ii) } N^{14} + n^1 \rightarrow C^{14} + p^1$$

The product, C^{14}, is beta active with a half-life of 5600 y and is widely used as a radioactive tracer, particularly in organic and biological systems. This (n,p) reaction is the only convenient way of producing C^{14}, since C^{13}, from which C^{14} could be produced by an (n,γ) reaction, has a very low capture cross-section and occurs in very low abundance in natural carbon. To obtain the same yield of C^{14} as in the irradiation of 1 g of nitrogen, 300 kg of carbon would be required.

(d) *Fission.* The compound nucleus formed when a neutron is captured by certain of the nuclei of heavy elements, instead of emitting a particle or gamma photon, splits into two roughly equal parts. This physical process was, in fact, discovered by chemists using established radio-chemical techniques. In 1938 Hahn and Strassman were investigating the nature of the radio-active products formed by the irradiation of uranium with low energy neutrons. Experience with a wide range of elements had shown that (n,γ) capture resulted in the formation of a beta-emitting nuclide which decayed to an element of higher atomic number. The irradiation of uranium did give beta-active products, and these could be separated from the parent uranium by co-precipitation with barium chloride. Four distinct isotopes, with half-lives ranging from about a minute to 300 h, were identified in the precipitate. These activities, being co-precipitated with barium, could be attributed only to isotopes of either barium or radium. As no physical process then known could account for the production of barium by neutron capture in uranium, Hahn and Strassmann assumed that the activity was due to isotopes of radium formed from the uranium by the successive emission of two alpha particles. Their attempts to separate the " radium " isotopes from the barium carrier by well-established radiochemical methods were, however, completely unsuccessful. By most careful experiments they were able to show that traces of added radium could be separated from the mixture of barium and " radium " isotopes, and that the latter always remained with the barium carrier. This established beyond doubt that radioactive barium was one of the products of the neutron irradiation of uranium. This was confirmed by the identification of lanthanum in the decay products of the " radium " isotopes. The results were published early in 1939; the conclusion was drawn that the nucleus of uranium had split into two roughly equal parts under the neutron irradiation. Almost immediately the physicists Meitner and Frisch showed that fission of uranium could be explained on the physical theories then current, and that the process should be accompanied by a large release of energy. Physicists quickly devised experiments to detect the energetic products of fission, and confirmatory results were reported from several centres in the United States. Within a few months many workers had demonstrated that fission resulted in not only barium and lanthanum, but also in other elements of medium atomic weight, such as bromine, krypton, strontium, molybdenum, rubidium and iodine,

NEUTRON CROSS-SECTIONS

It is convenient to assign to each nucleus a cross-section for each of the processes which may occur during the interaction with neutrons; scattering, (n,γ), (n,p) or (n,a) capture, or fission. These can be measured either by observing directly the attenuation and scattering of a beam of neutrons of known intensity and energy or by measurement of the activity, and hence the number of atoms produced, of the radioactive products of neutron capture. Some values of cross-sections of interest in the atomic energy industry are collected in Table 3.

TABLE 3.—NEUTRON CROSS-SECTIONS OF CERTAIN ELEMENTS

Element	Cross-section (barns, 1 barn = 10^{-24} cm^2)			
	At 0·025 eV		At 1·0 MeV	
	(n,γ) capture	Scattering	(n,γ) capture	Scattering
H	0·32	20–80	$< 10^{-4}$	4
D_2O	9.10^{-4}	15	$< 10^{-4}$	14
Be	9.10^{-3}	6·9	$< 10^{-4}$	3
B	718	3·8	$< 10^{-4}$	2
C	$4·5.10^{-3}$	4·8	$< 10^{-4}$	2·5
Al	0·22	1·6	4.10^{-4}	3
Fe	2·5	11·0	8.10^{-3}	3
Zr	0·4	8·4	3.10^{-2}	7
Cd	3500	6·5	10^{-1}	7

At low neutron energies the range of values covered by the (n,γ) cross-section is very large, more than six powers of ten. At high neutron energies, (n,γ) capture is of minor importance compared with scattering, and the values of the cross-sections increase quite regularly with increasing atomic number of the target nucleus.

NATURE OF THE FISSION PROCESS

After the discovery of fission by chemical techniques the process was intensively investigated by physical methods. These soon demonstrated two important aspects of the process of fission, namely, that the fission-fragments possess very high kinetic energy and that in the act of fission more than one neutron is released; this immediately opened up the possibility of achieving a self-sustaining fission process. These two facts alone are the foundations upon which the whole concept of atomic energy is based.

Hahn irradiated natural uranium, which consists of 99·3 per cent of U^{238} and 0·7 per cent of U^{235}; only the U^{235} isotope is fissile to neutrons of thermal energies. The beta activities of U^{239} and Np^{239}, produced by (n,γ) capture in U^{238} were too weak to be detectable in the presence of the relatively large fission-product activity. Although U^{235} is the only naturally-occurring isotope which undergoes fission with slow neutrons, many artificially-produced isotopes do so. This property, however, seems to be confined to those isotopes for which the sum of the atomic number and atomic weight is odd.

Fission is induced in a wider range of nuclei by particles of high energy. The threshold energy above which fission takes place depends upon the nucleus and particle involved. Thus both Th^{232} and U^{238} undergo fission with neutrons of energy greater than about 1 MeV, and most elements heavier than bismuth undergo fission on bombardment with protons, provided the proton energy is sufficiently high to overcome the coulomb repulsion of the nucleus. However, because of the large energy required to accelerate the protons these processes are of academic interest only and have no practical significance as energy sources.

As it is intended to achieve a self-sustaining reaction based on fission by neutrons, it is essential to know accurately the number of neutrons released per fission, a quantity usually denoted by the symbol ν. This has been determined, as a function of neutron energy, by direct counting of the neutrons released on fission for the isotopes U^{235}, U^{233} and Pu^{239}. Not every neutron capture leads to fission; a certain proportion result in an (n,γ) capture, even in the fissile nucleus, so that the number of neutrons emitted per

TABLE 4.—NUCLEAR PROPERTIES OF THE FISSILE ISOTOPES

Fissile Material	Neutron energy: 0·025 eV				1·0 MeV			
	σ_f barns	σ_c barns	ν	η	σ_f barns	σ_c barns	ν	η
Pu^{239}	740	290	2·9	2·1	2	~0·1	3	3
U^{233}	530	60	2·5	2·3	2	~0·1	2·5	2·5
U^{235}	580	110	2·5	2·1	1·3	~0·1	2·5	2·5
U^{238}	0	2·7	0	0	0·5*	~0·15	—	—
Natural uranium	3·9	3·5	2·5	1·3	0·01	~0·15	2·5	0·2

* At 2 MeV (threshold ~ 1·5 MeV).

capture in fissile material (denoted by η) is less than ν by the ratio of the fission cross-section σ_f to the total cross-section ($\sigma_f + \sigma_c$). The values of cross-sections, ν and η for fission of the nuclides of practical importance are shown in Table 4.

As with the elements listed in Table 3, the cross-sections are much smaller at the higher neutron energy, especially those for (n,γ) capture. This implies that at high neutron energies a greater proportion of the neutrons captured by the fissile material cause fission, and fewer are wasted by (n,γ) capture in the fuel.

ENERGY OF THE FISSION PROCESS

The total energy released per fission is about 195 MeV, this being determined experimentally by calorimetry and by calculation from the mass loss on fission using the Einstein relation $E = \text{Mc}^2$. This energy is distributed between the fission fragments, the gamma energy emitted at the instant of fission and the decay energy of the radioactive fission products in the proportions shown in Table 5.

TABLE 5.—DISTRIBUTION OF ENERGY RELEASED PER FISSION OF U²³⁵ BY THERMAL NEUTRONS

	Energy (MeV)	Per Cent of Total
1. Kinetic energy of the fission fragments	162	83·0
2. Kinetic energy of the neutrons	6	3·1
3. Gamma energy released at instant of fission	6	3·1
4. Gamma decay energy of fission products	5	2·6
5. Beta decay energy of fission products	5	2·6
6. Neutrino energy associated with the β-decay	11	5·6
TOTAL	195	100

This energy of 195 000 000 eV which is released on fission of one atom of uranium may be compared with the 4 eV released on the combustion of one atom of carbon.

The fission fragments interact strongly with matter, so that in a condensed phase they are quickly brought to rest, their energy ultimately appearing as heat. The beta and gamma decay energy of the fission products is released over a period of time, but because many of the fission products have short half-lives much of this energy is released while the fuel is in the reactor. The beta particles, like the fission fragments, are soon stopped by solid materials and their energy appears as heat, but the proportion of the gamma energy appearing as useful heat within the reactor depends upon the design of the core, the remainder being absorbed in the shielding surrounding the reactor. Neutrinos, which are associated with beta decay, interact only slightly with matter and therefore contribute nothing towards the useful heat output. Of the 195 MeV available per fission, therefore, roughly 173 to 184 MeV, or 89 to 94 per cent of the total is available for the production of useful heat. This corresponds to the production of roughly 10^{10} kcal or 10 million kWh of heat per pound of fissile material consumed, compared with 3000 kcal or 3 kWh/lb of coal burnt.

It is the object of the reactor engineer to design systems in which fissile material is aggregated in such a way in the reactor core that the fission of uranium or plutonium is accomplished smoothly, safely and efficiently, and the chemist has an important part to play in helping to achieve this end.

THE ACHIEVEMENT OF A SELF-SUSTAINING NUCLEAR REACTION

The fission process can only be maintained if, on the average, one of the neutrons released on fission is captured by a second atom of fissile material and so on; all methods of calculating the critical size of a given system, however complex, are based on this balance. A neutron, born by fission in an assembly of fissile material can suffer one of three fates; it may (a) escape from the system and be lost; (b) be captured by any material in the core, excluding fissile material; or (c) be captured by the fissile element, so furthering the chain reaction. A self-sustaining chain reaction can be achieved in either of the two main types of reactor, which are classified according to the average energy of the neutrons in the system.

FAST REACTORS

The neutrons of 2 MeV produced on fission are capable of initiating further fissions in the reactor core but, as may be seen from Table 4, the cross-sections of the fissile isotopes to neutrons of this energy is small. Consequently a high density of pure fissile material is required so that sufficient of the neutrons are captured by the fissile material before they escape from the system. Natural uranium, containing only 0·7 per cent of the fissile isotope U^{235} can never be used as a fuel for a fast reactor, because of the excessive absorption in U^{238} (cf. the value of η, Table 4). However by separating out the U^{235} by gaseous diffusion of uranium hexa-fluoride or by using plutonium 239 produced as a by-product in other reactor systems, a self-sustaining chain reaction can be achieved, as in the low-power reactors Zephyr and Zeus at Harwell. Zephyr is an experimental reactor, running at a power of only a few watts, built to study the properties of a fast critical assembly. The core, about the size of a top-hat, consists of a close array of pencil-like rods of plutonium 239, canned in nickel to avoid oxidation, surrounded by a reflector of uranium 238. Neutron capture occurs in the U^{238}, leading to the production of Pu^{239}. It has been demonstrated that more Pu^{239} is produced in the blanket than is consumed in the core.

Zeus is a mock-up of the experimental power reactor which is being built at Dounreay. The fuel elements are rods of U^{235}, and the reflector consist of an assembly of rods of U^{238} surrounding the core. To simulate the liquid sodium coolant which is to be used in the power reactor, rods of magnesium are interspersed with the fuel and reflector rods. Because of the increased spacing of the fuel, the core of Zeus is larger than the core of Zephyr, being a right cylinder two feet high.

THERMAL REACTORS

An effective means of modifying the nature of the reactor core is to make use of the fact that neutrons of low energy interact much more strongly with matter than do those of high energy. The required slowing-down or *moderation* of the 2 MeV fission neutrons is achieved by introducing into the core a large amount of an element of low atomic weight and high scattering cross-section, the moderator. It is essential that the absorption cross-section of the moderator be low so that few neutrons are lost in the process of slowing down. Suitable moderators are: (i) hydrogen, in the form of water or hydrocarbon; (ii) deuterium, in the form of heavy water, D_2O; (iii) graphite; (iv) beryllium, as metal or oxide.

In spite of the more efficient capture of thermal neutrons by the fuel, a homogeneous mixture of natural uranium and a moderator can never be made critical, since the abundant U^{238} has a very high capture cross-section for neutrons of certain energies between 2 MeV and thermal energies, known as *resonance peaks*. Consequently insufficient neutrons to maintain a chain reaction reach thermal energies. This difficulty is overcome by concentrating the natural uranium, usually as the metal or oxide, in *fuel elements* which are distributed throughout the moderator. The fast fission neutrons leave the fuel element (some, however, initiate a small number of fissions before they can escape) and enter the moderator where a substantial proportion are slowed down below the energy of resonance capture in U^{238} before re-entering a fuel element and initiating further fission. Some of the thermal neutrons are, however, captured in structural materials in the core rather than in the fissile material.

During the process of slowing down, and also while diffusing in the moderator a proportion of the neutrons escape from the core, the smaller the core the greater the number lost. This loss can be cut down by surrounding the core with a *reflector*, often of the same material as the moderator, the purpose of which, as the name suggests, is to reflect back some of the escaping neutrons and so make them available for sustaining the chain reaction in the core. Using natural uranium, the margin of spare neutrons is small, so that emphasis is placed on the need to keep down to an absolute minimum the absorption of neutrons by structural materials and by impurities in the moderator and fuel. Furthermore, the number of neutrons escaping from the core must be small, so that an essential feature of a natural uranium system is a large physical size. The neutron cycle in a graphite-moderated natural-uranium reactor such as BEPO is shown diagrammatically in Fig. 4.

Enriched Fuel System

If the proportion of U^{235} in the uranium fuel is increased the effect of capture in U^{238} is reduced and therefore more neutrons are available for other purposes, such as capture in necessary structural material or in a more efficient coolant, or escape from the core, thus permitting a smaller system to be constructed.

Breeding and Conversion

As seen in Fig. 4, in a reactor fuelled with natural uranium about 36 per cent of the neutrons released on fission are captured by U^{238} to give U^{239}, which decays to the fissile isotope plutonium 239. Thus, neutrons lost by capture in U^{238} are not entirely wasted, but go to produce a new nuclear fuel. In practice, graphite-moderated reactors of the Calder-Hall type produce between 0·6 and 0·7 atoms of Pu^{239} for every atom of U^{235} destroyed, so that 60 or 70 per cent of the U^{235} undergoing fission may be regarded as having been *converted* to Pu^{239} by capture in the *fertile* U^{238}. Some of the Pu^{239} undergoes fission while the fuel element remains in the reactor, so contributing to the useful heat output, and the remainder can be separated out chemically (Chapter 5) and used either alone as fuel in a fast reactor or in admixture with uranium which

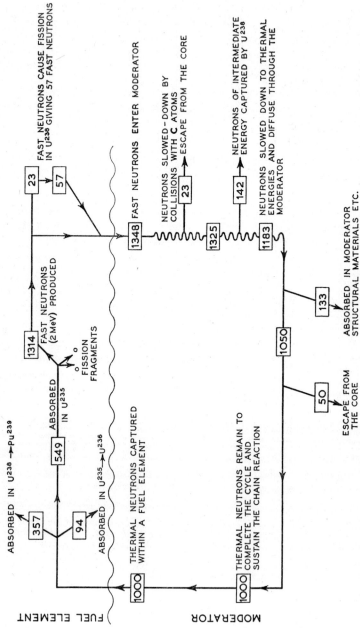

Fig. 4. Neutron Cycle of a Typical Natural Uranium, Graphite Moderated Reactor (e.g. BEPO)

has been depleted in U^{235} as fresh fuel for the thermal reactor.

From the point of view of the nuclear physicist the mixing of fissile and fertile materials is not the best way of achieving a high degree of conversion. Much better conversion can be achieved by concentrating the pure fissile material in the reactor core (either thermal or fast) and surrounding this by a *blanket* of fertile material which may be either U^{238} or Th^{232}, giving Pu^{239} and U^{233}, respectively. In this way a conversion exceeding unity can be achieved, so that the system *breeds* more fissile material than is consumed. The degree of conversion attainable depends largely on the number of neutrons emitted on fission per neutron captured—the η-value—which depends not only upon the fissile isotope considered but also upon the neutron energy at which fission takes place. The maximum conversion possible for various combinations is listed in Table 6; the conversion attainable in practice is usually somewhat less than these values due to the loss of neutrons by capture in other materials.

TABLE 6.—POSSIBLE CONVERSION FACTORS FOR VARIOUS
REACTOR SYSTEMS

Fissile Isotope	Fertile Isotope	Product	Moderator	Reactor Type	Conversion
U^{235}	Th^{232}	U^{233}	D_2O	Thermal	1·1
U^{233}	Th^{232}	U^{233}	D_2O	Thermal	1·3
U^{233}	Th^{232}	U^{233}	None	Fast	1·5
Pu^{239}	U^{239}	Pu^{239}	None	Fast	1·9

These values of conversion which might be obtained for reactors using enriched fuels are much higher than the values of 0·6 to 0·7 obtained in a natural uranium reactor.

Conversion or, better still, breeding, is of considerable economic importance, since the fuel for nuclear reactors is then not limited to the relatively scarce U^{235}, but includes the much more abundant materials U^{238} and Th^{232}. The high conversion attainable in fast reactors is an important reason for the present great interest in their development.

ESSENTIAL FEATURES OF ANY REACTOR SYSTEM

Whatever the type of power reactor system, certain essential conditions must be fulfilled. These are:

(*a*) The geometrical arrangement of fuel and moderator (in a thermal reactor) must be such that a chain reaction proceeds within the core.

(*b*) Some form of fluid coolant (liquid or gas) must be provided to remove the fission heat from the core.

(*c*) Considerable shielding round the core is necessary to protect personnel from the intense neutron and gamma-radiations arising in the core.

(*d*) Perfect containment of the radioactive fission products within the reactor is essential, either in the form of a hermetically sealed *can* surrounding a solid fuel element, or in absolute leak-tightness in systems with fluid fuel.

(*e*) Some means of control is required so that the neutron balance within the core can be adjusted so that the reactor power is held constant. This is usually achieved either by varying the amount of fuel in the core, or by introducing absorbing materials into the core.

POSSIBLE TYPES OF REACTOR SYSTEM

The units which go to make up a reactor are the fuel, moderator and coolant, and combining these with the variations of blanket and of fuel system leads to a wide variety of possible reactor types, as illustrated in Table 7, which is by no means exhaustive.

TABLE 7.—VARIABLES IN THE DESIGN OF A REACTOR

Fuel	Form of Fuel	Moderator	Coolant	Construction
1. Natural uranium 2. Natural uranium slightly enriched in U^{235} 3. Pure U^{235} 4. U^{233} 5. Pu^{239}	1. Metal rods 2. Oxide compacts 3. Metal or compound dissolved or suspended in suitable solvent (e.g. D_2O, liquid bismuth, fluoride melts)	1. None (fast reactor) 2. Graphite 3. Beryllium 4. Beryllia 5. H_2O 6. D_2O 7. Organic material	1. Gas (CO_2, H_2, etc.) 2. H_2O 3. D_2O 4. Organic liquid 5. Liquid metal (e.g. Na, Bi) 6. Fused salt (e.g. NaOH) 7. The fuel itself, if liquid (cf. col. 2, item 3)	1. With or without blanket of fertile material 2. Fuel and moderator intimately mixed (homogeneous) or segregated (heterogeneous)

Many of the possible combinations may be eliminated *a priori*; for example, water cannot be used as a coolant for a fast reactor, and a homogeneous mixture of natural uranium and graphite could, on nuclear grounds, never be made critical. Even after this elimination the number of remaining combinations is still imposing.

The final decision as to the acceptability of a certain reactor type for power production can only be made after an assessment of the cost of the power sent out. For reactors intended for electricity generation this must be competitive with power generated by conventional means. This should be achieved in the gas-cooled, graphite-moderated, natural uranium reactors which are under construction, but development work is still proceeding on more advanced types of reactors; the chemical problems associated with some of these will be discussed in later chapters. It should be noted that whereas reactors fuelled with natural uranium can be made economic with a low conversion factor, the cost of fully-enriched fuel is so high that in order to achieve economic power about 100 per cent conversion is required, so that the fuel burnt in the core can be replaced by that produced in the fertile blanket. Furthermore the chemical processes used to extract the small amount of fissile material produced in the fertile blanket should, if possible, be simple and cheap.

PROPERTIES OF THE FISSION PRODUCTS

NATURE OF THE FISSION FRAGMENTS AT THE MOMENT OF FISSION

The fission energy carried by the two fission fragments is unevenly distributed between the light and heavy fragments; by the observation of the ionisation tracks left by fission fragments in cloud-chambers and photographic plates the average properties of the two light and heavy fragments has been ascertained, and these are summarised in Table 8, together with the corresponding values for a 5 MeV alpha particle.

TABLE 8.—AVERAGE PROPERTIES OF THE FISSION FRAGMENTS

	Light Fragment	Heavy Fragment	5 MeV Alpha particle
Atomic mass (mean)	95	139	4
Atomic number (mean)	38 (Sr)	54 (Xe)	2
Initial charge	+ 20e	+ 22e	+ 2e
Kinetic energy (MeV)	97	65	5
Initial velocity (cm/sec)	$1 \cdot 4.10^9$	$0 \cdot 93.10^9$	$1 \cdot 3.10^9$
Range in air at atmospheric pressure (mm)	25	19	35
Initial rate of energy-loss in air (MeV/mm)	7·8	8	0·1

The average range has been measured in various materials by

observing the penetration of thin films by fission fragments originating in a foil of irradiated fissile material, and some typical values are presented in Table 9.

TABLE 9.—AVERAGE RANGE OF FISSION FRAGMENTS IN VARIOUS MATERIALS

Material	Al	Cu	U	U_3O_8	UO_2	H_2O
Mean range of fission fragment (in micron = 10^{-3} mm)	10	4	5	9	8	25

The high energy and charge of the fission fragments implies that strong interaction with the surrounding medium occurs over most of the path of the fragment; this is demonstrated by the high specific ionisation in air. This interaction has strong repercussions on the efficient running of a reactor, for during operation the fuel is continually being subjected to this intense bombardment.

It has been estimated that in a metallic fuel element the temperature attained for a very short time over the path of the fission fragment is of the order of several thousand degrees centrigrade. This estimate is borne out by recently published Russian work in which the atoms which evaporated from a cold thin film of fissile material during neutron irradiation were condensed upon a receiving foil and the quantity estimated by counting techniques. The number of atoms evaporated per fission in the sample was found to be, for plutonium metal 3500, uranium metal 1200 and uranium oxide 24.

RELATIVE YIELDS OF THE FISSION PRODUCTS

The act of fission is a statistical process, and the products resulting from the irradiation of a fissile material by thermal neutrons contain a distribution of elements with masses lying between 72 and 161. This range covers the elements zinc to dysprosium in the Periodic Table. Furthermore the act of fission is not symmetrical, in that instead of giving fission fragments of roughly equal mass, the most probable masses lie at about 93 and 140. The general shape of the curve of fission yield against mass number for the fission of U^{235} by thermal neutrons is shown in Fig. 5. The curve is symmetrical about a mass number of 117. The total mass number before fission (including the neutron which initiated the process) was 236, and after fission 234. The difference of two represents an approximate value for the number of neutrons emitted per fission (the quantity ν).

Curves of similar shape are observed for the slow-neutron fission of Pu^{239} and U^{233}, the curves being shifted slightly to account for the variation in mass number of the fissile isotopes. As the energy of the incident neutron is increased, symmetrical fission becomes more probable; at sufficiently high energies the minimum disappears and a single-humped curve is obtained.

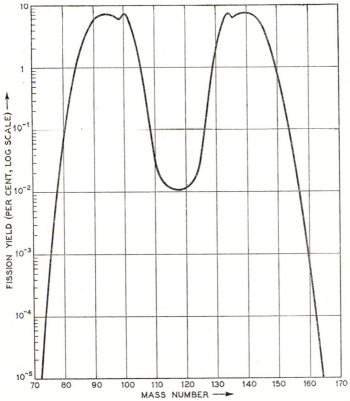

FIG. 5.   YIELDS OF THE VARIOUS MASS NUMBERS PRODUCED IN THE FISSION OF U^{235} BY THERMAL NEUTRONS

At present there is no adequate physical theory to explain the process of fission, and precise measurement of the yields of the various mass numbers is of great value in providing experimental data for the theoretical physicist. This is particularly true when the measurements are of sufficient precision to show up slight irregularities in the yield-mass curve, such as are shown near the peaks in Fig. 5.

A knowledge of the absolute yields of the fission products is also of practical importance. Some of the fission products, such as Xe^{135} and Sm^{149}, have very high neutron capture cross-sections, and so interfere with the neutron balance within the reactor. Others, such as Sr^{90}, are serious biological hazards. With a knowledge of the yield in fission, the half-life for radioactive decay and the capture cross-section of each of the fission products, it is possible to calculate, for any given times of irradiation and decay, the amount present, the activity and the total neutron absorption for any fission product.

Determination of the Mass-yield Curve

If we consider any one mode of fission of U^{235}, two sets of fission fragments are produced, such that the sum of their mass numbers is 234. These fragments, as we have seen, are highly charged, but they rapidly neutralise this charge by capture of electrons from the surrounding medium. The resulting fission products are unstable, and emit beta particles which are often accompanied by one or more quanta of gamma-radiation, until a stable nuclide is reached. As a typical example we may take the complementary mass-chains 90 and 144 resulting from the fission of U^{235}.

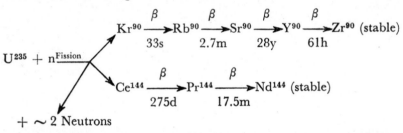

In any one mass-chain the process of fission leads directly not only to the first member of the chain. Other members of the chain are produced directly in fission to a greater or less extent. If it were possible to measure very soon after the instant of fission the amount of each nuclide produced, an *independent* fission yield could be determined. This is possible in only a limited number of cases, owing to the very short half-lives which are usually found in nuclides at the beginning of the decay chain. In practice the *cumulative* yield for all the nuclides of a given mass number is determined, and this is usually the quantity which is of interest. Knowing the cumulative fission yield it is possible to calculate theoretically the independent fission yields of each nuclide, when this is required.

Experimentally, the cumulative fission yield for a mass chain can be determined either by physical methods, such as mass-spectrometry, or by chemical methods. The latter involves the radio-chemical separation of a particular element of suitable half-life and determining the amount present by beta or gamma counting. Normally the absolute yield is not determined, but rather the relative yields of a series of elements. These are plotted as a mass-yield curve, and the yields are normalised by equating the integral of the curve to 200 per cent, the total yield of fission products. For the determination of, for example, Cs^{137} a small sample of fissile material is irradiated in a nuclear reactor, usually for a period of minutes or hours, and the sample is dissolved in acid and made up to a known volume. To an aliquot of this solution a known amount of a caesium salt is added. This acts as a carrier for the minute amount of radio-active caesium which is present. Caesium perchlorate is precipi-tated by the addition of perchloric acid followed by ice-cold alcohol and, after washing with alcohol, the precipitate is dissolved in water. The solution is then scavenged of any residual contamination by a triple precipitation of ferric hydroxide, the caesium remaining in solution. Finally caesium chloroplatinate is precipitated, mounted on a counting tray and dried. These separation and purification procedures need not be quantitative, for the resulting sample is weighed, and a correction factor for any losses is applied. The activity of the sample is determined by standard counting techniques, and, when possible, the radiochemical purity is checked by observing the activity over a period of time in order to determine the half-life of the sample.

RADIOACTIVITY OF THE FISSION PRODUCTS

On the average, each of the pairs of mass-chains which are pro-duced in fission contain between five and six radioactive nuclides of half-life less than a few weeks. When a reactor has been operating at power for this length of time most of these short-lived isotopes are decaying as rapidly as they are being produced in fission. Every fission, therefore, results in about five beta decays. Each MW of reactor power corresponds to a fission rate of 3.10^{16} fissions per second or about 10^6 curies of fissions. The radioactivity associated with a reactor heat output of one MW is therefore of the order of 15.10^{18} beta decays per second, or about 5.10^6 curies of beta activity, with roughly the same amount of associated gamma activity. Because of the large number of very short-lived isotopes (with half-lives of minutes or less) among the fission products, the activity falls off very rapidly on shut-down of the reactor, as shown in Fig. 6.

The heat output associated with the fission-product activity is

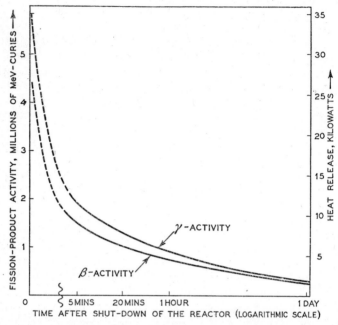

FIG. 6. DECAY, AFTER SHUT-DOWN, OF THE FISSION-PRODUCT ACTIVITY RESULTING FROM THE PRODUCTION OF ONE MEGAWATT OF HEAT BY FISSION.

also shown in the figure. All of the energy of beta decay is deposited locally. As the gamma rays are more penetrating than beta particles, only a proportion of the gamma decay energy is deposited within the fuel. The actual heat release in the fuel on shut-down of a reactor therefore lies somewhere between the value for the beta energy alone and the total beta plus gamma energy. In any event, the heat release in the early stages of decay is a significant proportion of the total reactor power.

EFFECT OF THE FISSION PRODUCTS ON THE OPERATION OF A NUCLEAR REACTOR

The effect of fission fragments on the operation of a reactor is two-fold. Firstly, the energetic fission fragments cause physical damage to the fuel, at a rate and in a manner which depends upon the nature of the fuel system. Secondly, all of the fission products absorb neutrons to a greater or less extent, so that as the operating time of the reactor increases an increasing number of neutrons is captured by the fission-product *poisons*. This increasing neutron absorption, coupled with the consumption of the fissile material, upsets the neutron balance within the reactor, and so it would eventually become subcritical and the chain reaction would cease.

RADIATION DAMAGE OF THE FUEL (See Plate VIII)

In fuel elements of metallic uranium severe distortion occurs after prolonged irradiation and this may result in the rupture of the fuel-element can. This effect has been likened to the wrinkling and warping which occurs when uranium undergoes repeated thermal cycling. Typically, a single crystal of alpha uranium, initially cylindrical in shape, becomes elongated and oval in cross section. As yet there is no simple explanation for this effect, but it seems to be associated with the very localised melting of the metal along the paths of the fission fragments. Samples of uranium also increase in volume on irradiation. This increase is attributed to the build-up of the inert gas fission products (Xe, Kr). These diffuse through the metal and collect under very high pressure in the intercrystalline pores. This mechanism is substantiated by the observation that the increase in the crystal dimensions of uranium on irradiation, as determined by X-ray diffraction techniques, is insufficient to account for the observed decrease in density.

Although in a natural-uranium reactor sufficiently long operating times can be achieved before reaching the point where there is danger of rupture of the can, it is obviously economically desirable to prolong the useful life of the fuel element as much as possible. This may be achieved with a metallic fuel element either by increasing the strength of the can or by suitably alloying the uranium to produce a tougher metal. Both of these methods, however, introduce further neutron absorbers into the reactor core, giving a poorer neutron economy. Alternative fuel-element materials may also be used, such as uranium oxide or carbide, but these introduce other problems, which will be discussed in Chapter 8. Another possibility is to suspend or dissolve the fissile material in some suitable liquid, such as water or a liquid metal. Again, difficulties are introduced, and these are discussed in Chapters 7 and 9.

FISSION-PRODUCT POISONING OF THERMAL REACTORS

As mentioned above, knowing the yields, half-lives and neutron absorption cross-sections for each fission product it is possible to calculate the amounts of each fission product present after a given irradiation time. Any radioactive nuclide in a chain may either capture a neutron, and so enter the next higher mass-chain, or may undergo radioactive decay. As an example we may take the important chain number 135:

$$\text{Fission} \rightarrow \text{Te}^{135} \xrightarrow{\beta} \text{I}^{135} \xrightarrow{\beta} \text{Xe}^{135} \xrightarrow{\beta} \text{Cs}^{135} \xrightarrow{\beta} \text{Ba}^{135} \text{ (stable)}$$

	Te^{135}	I^{135}	Xe^{135}	Cs^{135}	Ba^{135}
	2m.	6.7h.	9.2h.	2.1 10^{6}y.	
(n,γ)	(n,γ)	(n,γ)	$(n,\gamma)\ \sigma_c = 3.2 10^6$ barns.	$(n,\gamma)\ \sigma_c = 15$ barns.	$(n,\gamma)\ \sigma_c < 4$ barns.
	Te^{136}	I^{136}	Xe^{136}	Cs^{136}	Ba^{136}

The relative importance of radioactive decay and neutron capture depends for a given nuclide, on the neutron flux within the reactor; as this increases so does the neutron capture. For example, in BEPO (flux $\sim 10^{12}$n. cm^{-2} sec^{-1}) about 20 per cent of the Xe^{135} disappears by neutron capture to Xe^{136}, in DIDO (flux $\sim 10^{14}$n. cm^{-2} sec^{-1}) the figure rises to 98 per cent.

Several authors have been given estimates of the total neutron absorption of the fission products as a function of time for different neutron fluxes. An average figure for the cross-section of the fission products of U^{235} (excluding Xe^{135} and Sm^{149}) is 65 barns per fission. This implies that when a reactor has operated for such a time that 10 per cent of the initial U^{235} has undergone fission, then for every 100 neutrons captured in the remaining U^{235} one will be captured in the fission products (again excluding Xe^{135} and Sm^{149}). Typical values of the relative neutron absorption of the various elements, calculated as outlined above, are given in Table 10, together with their relative abundances.

TABLE 10.—RELATIVE ABUNDANCE AND NEUTRON ABSORPTION OF THE FISSION PRODUCTS

Element	Relative Neutron Absorption	Relative Abundance (per cent)
Rare gases	72	7
Samarium	14 ⎫	
Other rare earths	11 ⎬	70
Technetium	1	10
Caesium	0·5	4
Molybdenum	0·2	1
Remainder	1·3	8
	100	100

Xenon and samarium occupy a unique position among the fission-product poisons, for they are formed in high yield (5·6 and 1·3 per cent) and have very high capture cross-sections of 3 200 000 and 66 000 barns respectively. Samarium^{149} is stable, and only undergoes neutron capture. It therefore builds up in concentration during the operation of a nuclear reactor until its rate of formation is equal to its rate of neutron capture, and at this point, which is soon reached in a high flux reactor, its relative poisoning is equal to the yield of 1·3 per cent. Xenon, however, also undergoes radioactive decay, and only at higher neutron fluxes is the rate of neutron capture sufficiently high to compete with radioactive decay. The poisoning due to Xe^{135} is therefore small in a low-flux reactor and, like Sm^{149} poisoning, rises to a value equal to its yield (5·6 per cent) at higher fluxes.

The poison Xe^{135} is of further interest because it is produced by decay of a fairly long lived isotope I^{135}. During reactor operation the rate of formation of Xe^{135} by decay of I^{135} is equal to its rate of disappearance, which, in a high-flux reactor, is almost entirely due to neutron capture. When the reactor is shut-down, the rate of disappearance of Xe^{135} falls considerably to a value equal to its rate of radioactive decay. The concentration of Xe^{135} therefore rises to a maximum, at which point the concentration of I^{135} has fallen because it is no longer being produced in fission. After the maximum is passed the concentration of Xe^{135} falls off according to the normal exponential decay law with a half-life of 6·7 h. The maximum concentration of Xe^{135} is reached eleven hours after shut-down, and at this point the poisoning can be as high as 50 per cent. Unless a very large excess of reactivity has been built into the design, it is not possible to operate a reactor with this degree of poisoning. Unless the start-up of a reactor is therefore begun within a short time of shut-down, before the Xe^{135} concentration builds up, it may have to be delayed by as much as two days, after which time the Xe^{135} concentration has fallen to give a poisoning of about 5 per cent.

Means of Overcoming the Neutron Poisoning

The build-up of fission-product poisons, coupled with the consumption of uranium during operation would soon make a reactor become subcritical. This is avoided, in a reactor employing solid fuel elements, by loading more uranium into the reactor core than is necessary to achieve a critical system. This oversize core is prevented from becoming super-critical by the insertion of control-rods containing a strongly neutron-absorbing material, such as boron or cadmium. During operation of the reactor, when the poison level builds up, these rods are slowly withdrawn so that the total poisoning in the core remains constant.

In most thermal reactors employing enriched fuel (U^{235} or U^{233}), for economic reasons fresh fuel must be produced by neutron-capture in Th^{232} in a breeder blanket. The presence of neutron poisons in the core implies that fewer neutrons are available for breeding purposes. It is therefore essential to the economics of such reactors that the total fission-product poisoning be kept at a low level by continuously processing the fuel in some way. This is, in principle, feasible for liquid fuel reactors. The removal of Xe^{135} is then a comparatively easy problem, for, being gaseous it can be stripped out in a gas-liquid separator. Furthermore, it may be possible to remove the I^{135} continuously, so avoiding the build-up of poisoning after shut-down. The remaining fission products, however, require more elaborate processing systems, which are discussed in Chapters 7 and 9.

RADIOACTIVITY OF THE FUEL

The high heat output in the reactor fuel during the period immediately following shut-down implies that some cooling of the fuel is necessary during this period. This problem is particularly acute in reactors of high rating, in which a large amount of heat is extracted from a small amount of fuel. In reactors employing enriched uranium, such as a fast reactor or the research reactor DIDO the rating is about 1 MW/kg of fissile material. The heat output in the fuel soon after shut-down is therefore about 10 kW/kg (Fig. 6), and unless cooling is applied during this period melt-down of the fuel could occur. Provision has therefore to be made in the design of the reactor for emergency cooling, for example, by efficient convection, in the event of a failure in the main cooling system. Furthermore, when it is necessary to refuel a highly-rated reactor within a few hours of shut-down, cooling must be provided for the fuel elements during removal from the reactor and in subsequent storage.

In reactors fuelled with natural uranium the fission heat is produced at a much lower rating, of about 1 kW/kg. The problem of heat removal on shut-down is in this case small, as natural convection of the surrounding gas is sufficient to remove the fission-product heat.

The high gamma activity of the fission-products implies that irradiated fuel elements present a considerable health hazard. A feature of any reactor which employs solid fuel elements is therefore the heavy shielding associated with the unloading equipment. This usually takes the forms of a transport flask heavily shielded with lead, weighing many tons. In this flask fuel elements are transferred to a shielded storage facility where further decay of the activity takes place prior to subsequent processing.

POISONING OF A FAST REACTOR

At the high neutron energies of the fast reactor the cross-sections of the fission products are all low (cf. Table 3); none show the strong absorption which occurs in some isotopes at thermal neutron energies. The reactivity change due to fission-product poisoning in a fast reactor is therefore negligible compared with that arising from consumption of the fuel.

CHEMICAL EFFECTS OF FISSION FRAGMENTS

The fission fragments are interesting from the point of view of the radiation chemist because of the intense ionisation which they produce in passing through matter, the ionisation density being more than an order of magnitude greater than that of, for example, a 5 MeV alpha particle (cf. Table 8). In spite of this, the radiation chemistry of fission fragments has been investigated in only a limited number of systems. The decomposition of water is of importance in certain types of reactor, and is discussed in Chapter 7, and the radiation chemistry of gases is of interest as a possible means of producing chemicals, for example nitric acid, by fission-fragment irradiation in a power reactor. This aspect is discussed in Chapter 11.

Very little other work has been reported in the literature, and this has been confined mainly to studies of the decomposition of solid inorganic salts, such as potassium iodate or nitrate, when irradiated as an intimate mixture with, for example, uranium dioxide in a reactor. Several reasons might have contributed to bring about this apparent lack of interest. Relatively few research centres have had irradiation facilities in experimental reactors and irradiation space is often at a premium, and is therefore used for more pressing work, such as reactor development. Handling of the sample after the irradiation is often difficult owing to the radioactivity of the fission products, and special handling facilities are required. On the other hand, other sources of radiation are more generally available and, using these, the experiments are easier to perform and do not lead to high levels of radioactivity in the samples.

FURTHER READING

Stephenson. *Introduction to Nuclear Engineering.* McGraw-Hill Publishing Co. Ltd. (1954).

United Nations. Proceedings of the First International Conference on the Peaceful Uses of Atomic Energy, Geneva, 1955. Vol. 7, *Nuclear Chemistry and the Effects of Irradiation.*

Walton. " Fission Recoil and its Effects," *Progress in Nuclear Physics,* **6** (1957).

CHAPTER 4

THE NEW HEAVY ELEMENTS

THORIUM and uranium, the starting materials for a nuclear power programme, occur in nature and have been known since the early nineteenth and the late eighteenth centuries respectively. Protactinium was discovered in the form of a long-lived isotope in 1917, but its chemistry is not very well known, partly owing to its extreme rarity in nature; it occurs in uranium ores at a concentration of not more than 0·3 p.p.m. of uranium. All the transuranium elements may be regarded as man-made: very small quantities of neptunium and plutonium occur naturally, but the amounts are too small to provide a practicable source for extraction. Neptunium, for instance, has been isolated from Belgian Congo uranium ore concentrate; it occurs to the extent of only about two parts in 10^{12} parts of uranium, but such is the sensitivity of modern radiochemical techniques that separation and detection have been found possible.

PRODUCTION OF THE HEAVY ELEMENTS

There are two general methods available for the production of the elements of higher atomic weight than uranium, neutron irradiation in a nuclear reactor and bombardment by neutrons, or by heavier particles, at higher energies in particle accelerating machines. Both routes have certain disadvantages.

One of the limitations of the neutron capture reactions which are utilised in reactor irradiation is that an increase in atomic number is not a direct process, but can only take place as a result of the beta decay of the nuclei resulting from the neutron capture. As an example, we may take the production of one of the einsteinium isotopes from californium:

$$^{249}_{98}Cf \xrightarrow{(n,\,\gamma)} {}^{250}_{98}Cf \xrightarrow{(n,\,\gamma)} {}^{251}_{98}Cf \xrightarrow{(n,\,\gamma)} {}^{252}_{98}Cf \xrightarrow{(n,\,\gamma)} {}^{253}_{98}Cf \longrightarrow {}^{253}_{99}E + \beta \quad (1)$$

If one of the products of any proposed multiple neutron capture is an isotope which has a short half-life for alpha emission or for spontaneous fission, then the route becomes effectively blocked for proceeding higher up the series.

A nuclear reaction which involves the addition of a single neutron is often called *first order*. Addition of two neutrons is then a *second order* reaction, and so on. The concentration of an nth order product is usually dependent on the nth power of the integrated neutron flux, and so the advent of high flux reactors, such as the Materials Testing Reactor in Idaho, greatly increases the possibility of preparing detectable amounts of high-order products. Einsteinium and fermium, elements 99 and 100, were first identified in atomic debris from a thermonuclear explosion by teams working at the University of California Radiation Laboratory, at Los Alamos and at the Argonne National Laboratory in 1952. Uranium in this device had been subjected to an extremely high neutron flux of short duration, and this gave rise to multiple neutron captures of a much higher order than it had been possible to achieve previously by irradiation in nuclear reactors. Knowledge of the decay characteristics of the heavy-element isotopes gives rise to the conclusion that no less than 15 neutrons must have been added instantaneously to U^{238} to give U^{253} which decayed via seven intermediate beta emissions to E^{253}; similarly the primary step in the production of Fm^{255} must have been the instantaneous addition of 17 neutrons to the parent uranium. These high-order neutron capture reactions should be compared with the single-neutron capture to U^{239} which is normally as much as can be achieved in reactor fuel elements—the U^{239} has time to decay to Np^{239} before appreciable capture to give U^{240} can take place. Subsequent to the thermonuclear explosion work it was found possible to prepare several isotopes of einsteinium and fermium by prolonged irradiations in the MTR.

With the exception of einsteinium and fermium, each of the new transuranium elements has been isolated in the first instance as a result of cyclotron experiments with high energy particles. There is a greater possibility of control over the nuclear reactions than by reactor irradiations and it is possible to take much bigger jumps up the atomic weight series by bombardment with heavy ions such as those of carbon, nitrogen and oxygen. For comparison with the multiple neutron capture quoted above we may cite the reaction

$$^{238}_{92}U + [^{14}_{7}N]^{6+} \rightarrow {}^{247}_{99}E + 5n \qquad (2)$$

The limitations of the cyclotron method are the small beam currents of heavy ions which are obtainable, the limited range of energy over which a desired reaction will occur, and the small amounts of target material which it is possible to bombard.

The first of the transuranium elements to be isolated was neptunium, in May 1940. It was formed as the beta-emitting isotope Np^{239} by bombardment of uranium with a neutron beam in a 60-inch cyclotron:

$$^{238}_{92}U + ^{1}_{0}n \rightarrow ^{239}_{92}U + \gamma \tag{3}$$

$$^{239}_{92}U \rightarrow ^{239}_{93}Np + \beta \tag{4}$$

Earlier work had shown that fission-product atoms had sufficient energy to be ejected from a uranium foil under bombardment, and that they could be collected on inert foils placed nearby. McMillan and Abelson, in 1940, found that an activity with a 2·3 d half-life remained with the uranium foil. They were able to show by tracer techniques that the new product had chemical properties which distinguished it both from the original uranium and from the products of uranium fission. Subsequent work in the University of California was concerned with the products obtained by bombarding uranium with deuterons. This gave Np^{238} which decayed to Pu^{238} by beta emission. This first plutonium isotope was isolated towards the end of 1940 and its 100 y half-life enabled plutonium chemistry to be studied on a tracer scale. The more important, fissile, isotope Pu^{239} was isolated in 1941 from the neutron bombardment of uranium, reactions (3) and (4) occurred and were followed by reaction (5):

$$^{239}_{93}Np \rightarrow ^{239}_{94}Pu + \beta \tag{5}$$

Subsequent prolonged neutron bombardment of several hundred pounds of uranyl nitrate in the Berkeley and Washington University cyclotrons produced about $500\mu g$ of the new element by the end of 1942.

There was a gap of several years before plutonium became available in larger quantities by manufacture in the first reactors; by its subsequent re-irradiation in nuclear reactors it was possible to obtain americium in 1944:

$$^{239}_{94}Pu + ^{1}_{0}n \rightarrow ^{240}_{94}Pu + \gamma \tag{6}$$

$$^{240}_{94}Pu + ^{1}_{0}n \rightarrow ^{241}_{94}Pu + \gamma \tag{7}$$

$$^{241}_{94}Pu \rightarrow ^{241}_{95}Am + \beta \tag{8}$$

Similarly, cyclotron bombardment with high energy alpha particles gave curium,

$$^{239}_{94}Pu + ^{4}_{2}He \rightarrow ^{242}_{96}Cm + ^{1}_{0}n \tag{9}$$

The heaviest element to have been isolated at the time of writing is nobelium. A joint team of scientists from Sweden, Great Britain and the United States succeeded during 1957 in producing a minute quantity of an isotope emitting an 8·5 MeV alpha particle with a half-life of about 10 minutes. This material was produced by bombarding Cm^{244} with high energy ions of C^{13} and was tentatively assigned the mass number of 253, although confirmation has yet to be obtained. Seaborg and his collaborators claimed that experiments during 1958 did not confirm this discovery; they claimed instead to have synthesised 102^{254} (half-life 3s.) by bombardment of Cm^{244} with C^{12} ions. Table 11 summarises the historical aspects of the transuranium elements.

TABLE 11.—THE TRANSURANIUM ELEMENTS

Element	Atomic No.	First Isolation	First Isotope Produced	Authors	First Isolation as Pure Compound
Neptunium	93	1940	Np^{239}	McMillan and Abelson	1944
Plutonium	94	1940	Pu^{238}	Seaborg, McMillan, Wahl and Kennedy	1942
Americium	95	1944	Am^{241}	Seaborg, James, Morgan and Ghiorso	1945
Curium	96	1944	Cm^{242}	Seaborg, James and Ghiorso	1947
Berkelium	97	1949	Bk^{243}	Ghiorso, Seaborg and Thompson	1958
Californium	98	1950	Cf^{244}	Street, Ghiorso, Seaborg and Thompson	
Einsteinium	99	1952	E^{253}	} Ghiorso, Seaborg, et al.	
Fermium	100	1952	Fm^{255}		
Mendelevium	101	1955	Mv^{256}	Ghiorso, Harvey, Choppin, Thompson and Seaborg	
Nobelium (?)	102	1957	No^{253}	Atterling, Milsted, Fields and Friedman	

For the elements up to californium, it is now possible to produce weighable quantities by reactor irradiations. In fact substantial quantities of americium are produced in reactor fuel elements at present being processed at Windscale: this occurs by the successive capture of two neutrons by Pu^{239} in the reactor and beta decay of the Pu^{241} (reactions 6, 7, 8). The half-life of Am^{241} (468 y) is relatively short for an alpha-emitter, and this isotope contributes to the health hazard of the fission-product waste solutions from the extraction plant. The waste solution is not a convenient source of americium for experimental work owing to the very high levels of beta and gamma activities which are also present. It can be obtained much more easily from separated plutonium; the latter contains Pu^{241} produced during irradiation and, since its half-life is 13 y, the Am^{241} continues to grow after the plutonium has been separated. One kg of plutonium, consisting mainly of Pu^{239} but containing 1 per cent of Pu^{241}, will produce 0·5 g of Am^{241} in a year.

Plutonium reserved for the purpose of americium production in the laboratory is kept in an *americium cow* and the *milking* process of separating the americium (e.g. by ion exchange) is carried out after intervals of several months.

Curium is also best prepared in weighable amounts by reirradiation of separated americium in a reactor, rather than by attempting to separate it from highly irradiated uranium fuel elements. The production of any of the elements beyond curium in sufficient quantity to do more than tracer solution chemistry is a formidable task for several nuclear reasons, including spontaneous fission of the very heavy even-even isotopes. Moreover the half-lives of the isotopes which have been produced become progressively shorter with increasing atomic number. Thus, although seven isotopes of fermium have been isolated only one of them (Fm^{253}, half-life 4·5 d) has a half-life longer than one day.

Apart from the rapid disappearance of the isotopes with short half-lives, the handling of them in more than tracer amounts presents many practical difficulties if they are alpha-emitters. Even with Pu^{239} (half-life 24 000 y) the solution chemistry becomes somewhat modified by the action of the emitted alpha particles on water, and radiation damage soon becomes detectable in solid crystalline compounds. In the case of Cm^{242} (half-life 162 d) precipitates become difficult to separate from solutions owing to the intense decomposition of the water to give hydrogen and oxygen gas bubbles.

One other difficulty should be mentioned which applies to all tracer and ultramicrochemical work, but is often of greater importance in the handling of alpha-emitters. This is the gradual accumulation of impurities in the solutions due to the dissolution of glass containers. Milligram amounts of material may be leached from a glass surface of 100 cm^2 area by 3M–HCl. Work with intensely alpha-emitting solutions increases the rate of attack on the glass, and the products of the leaching process can become troublesome, for instance, in the accurate pulse-analysis of an alpha source by absorbing a considerable fraction of the energy of the emitted alpha particles. It is sometimes necessary to resort to other types of container materials, and a very strict control of the purity of all reagents is required.

THE ACTINIDE CONCEPT

As long ago as 1923 Bohr suggested that a series of elements might be possible in which electrons would populate the $5f$ energy levels; the properties of these elements would then resemble those of the lanthanides in which the $4f$ energy level is being filled. Several attempts were made subsequently to predict at which element this new series would begin, but the experimental evidence was rather scanty before the synthesis of the transuranium elements. Up to that time only four elements were known which could possibly form part of the new series of fourteen elements, and of these four only thorium and uranium had undergone any extensive chemical investigations. Moreover, as Bohr pointed out, the properties of the elements at the beginning of the new series would not necessarily be characteristic of f-shell behaviour, since the $5f$ electrons would be less effectively shielded than the $4f$ electrons of the lanthanides.

When McMillan and Abelson discovered the first transuranium element, neptunium, in 1940, they found that the product of their nuclear reaction could be distinguished from the parent uranium but nevertheless resembled this element more than rhenium, the element immediately above it in the Periodic Table when regarded as part of a heavy transition series. Moreover, at the end of the first year of Seaborg and Wahl's work with plutonium on a tracer scale it was evident that the latter also showed marked chemical similarity to neptunium and uranium, rather than to osmium.

By 1949, considerable chemical experience had been obtained with the new elements neptunium, plutonium, americium and curium and in that year Seaborg published a paper summarising the evidence for the *actinide concept* based on the marked similarities between these elements and the lanthanides. He postulated that the new series of elements could be written as if it began with actinium and that fourteen elements would follow which would form a series exactly analogous to the lanthanides—hence the adoption of the name *actinides* for the new series.

As an alternative to the actinide concept it may be supposed that the elements beyond actinium take part in a new transition series in which the successive unpaired electrons enter the $6d$ energy levels. By analogy, this should give quite different chemical properties from the supposition that the electrons are filling up the $5f$ energy levels. The experimental evidence will be discussed briefly in the following sections; it is quite obvious that the chemical properties of later members of the new series conform to the $5f$ hypothesis, but the evidence for the early members of the series (up to neptunium) is somewhat ambiguous.

OXIDATION STATES

The lanthanide elements exhibit a characteristic valency of three, with deviations from this occurring only in a few special cases. The first half of the actinide series is characterised by a marked tendency towards higher valency states, however; in fact all the elements uranium, neptunium, plutonium and americium show valencies of all values between three and six. The analogy between the two series becomes very strong in the later elements, where there is a marked tendency to favour the trivalent state in the actinides. The known oxidation states which exist in solution are shown in Table 12.

TABLE 12.—VALENCY STATES OF THE ACTINIDES AND THE LANTHANIDES IN SOLUTION

Actinium	3	Lanthanum	3
Thorium	4	Cerium	3, 4
Protactinium	4, 5	Praesiodymium	3, 4
Uranium	3, 4, 5, 6	Neodymium	3 (4)
Neptunium	3, 4, 5, 6	Promethium	3
Plutonium	3, 4, 5, 6	Samarium	2, 3
Americium	3, 4, 5, 6	Europium	2, 3
Curium	3	Gadolinium	3
Berkelium	3, 4	Terbium	3, 4
Californium	3	Dysprosium	3
Einsteinium	3	Holmium	3
Fermium	3	Erbium	3
Mendelevium	3	Thulium	3
Nobelium	3	Ytterbium	2, 3
(Element 103)		Lutecium	3

Despite the wider range of valencies of the actinides, there are many points of similarity between the members of the two groups. These are much more striking than the similarities of the heavier actinides to the corresponding d-shell transition elements if they are arranged immediately underneath the latter in the Periodic Table. The resemblance to the lanthanides is particularly strong in the middle of the series: curium and gadolinium both appear to occur only in the trivalent state in solution, and the tendency to form a stable f^7 configuration leads to the additional valency of four in berkelium and terbium. This evidence is not completely conclusive, however. Owing to the intense alpha emission of Cm^{242}, any curium solution rapidly acquires reducing properties due to decomposition of the water with hydrogen peroxide formation. The latter may mask any tendency for curium to form a higher valency state, but the addition of strong oxidising agents has given no evidence of oxidation. The wider use of high flux nuclear reactors will give a possibility of forming curium isotopes of a higher mass than Cm^{242}; in general, these are longer lived, less intensely radioactive, and would introduce less complications into the chemical studies. It has already proved possible to produce mg quantities of Cm^{244} by multiple neutron capture in plutonium irradiated in the Materials Testing Reactor in Idaho. The availability of substantial amounts of the longer-lived isotopes should be particularly valuable in enabling a more thorough search to be made for the possible existence of higher valency states of curium.

The penta- and hexavalencies shown by some of the actinides are not paralleled in the lanthanide series. Compounds of the type MX_5 and MX_6 are known within the actinide series, but the most stable compounds of these valencies (in fact, the only forms which exist in aqueous solution) involve the oxygenated ions XO_2^+ and XO_2^{2+}.

The absorption spectra of aqueous solutions of the lanthanide and the corresponding actinide elements in the same valence states show many similarities and this is particularly striking for the pair Eu(III) and Am(III). Moreover similarities also exist along the actinide series; thus the isoelectronic groups of ions NpO_2^+ and PuO_2^{2+}, and U III, Np IV and Pu V, show very closely related spectra.

CRYSTALLOGRAPHY

Extensive crystallographic studies of compounds of the actinide elements have been carried up the series only as far as americium. There is a very limited amount of data on curium, but as in the study of curium solution chemistry complications arise due to the intense radioactivity of the most readily available isotope Cm^{242}. Many series of isomorphous compounds have been shown to exist in these elements. Thus the following isostructural series may be quoted as examples:

ThF_4—UF_4—NpF_4—PuF_4—AmF_4—CmF_4 (monoclinic)
ThO_2—UO_2—NpO_2—PuO_2—AmO_2—CmO_2 (face-centred cubic)
   UF_3—NpF_3—PuF_3—AmF_3—CmF_3 (hexagonal)
   UF_6—NpF_6—PuF_6 (orthorhombic)
   UO_2^{2+}—NpO_2^{2+}—PuO_2^{2+}

These isostructural series form some of the most convincing evidence in favour of the actinide concept. Even here, however, there are exceptions to a close analogy with the lanthanides; for instance it has recently been found possible to synthesise some curium compounds (CmO_2 and CmF_4) in a higher valency state than three.

MAGNETOCHEMISTRY

Measurements of magnetic susceptibility can give evidence on the energy levels occupied by any unpaired electrons which an ion may possess. In earlier parts of the Periodic Table, the magnetic susceptibility of ions containing d- and f-level electrons has been found to differ quite markedly. The application of such measurements to compounds of the heaviest elements should give information on their electron configurations and may indicate which element is the first one to exhibit f-electron properties.

The measured magnetic moment of an ion includes contributions from both the orbital and the spin motions of the unpaired electrons. Provided that Hund's rule applies to these heaviest elements, then each additional electron in excess of the radon core will enter the most stable electron shell so as to maintain the maximum multiplicity of the spin quantum number. For the $5f$ shell, therefore, the electrons should not begin pairing until seven sub-states have been filled; for the $6d$ shell, five sub-states. Consequently we should be able to compare the early members of the actinide series with the members of the first halves of the lanthanide and the d-shell transition series. For the d-shell elements, such as scandium to zinc, good agreement with the observed magnetic susceptibilities is obtained by assuming that the orbital electron contribution is repressed by crystal electric field interactions or by exchange forces between neighbouring ions in the crystal lattice. The magnetic susceptibility then has *spin-only* values. On the other hand, the unpaired electrons in the $4f$ shell of the lanthanides are shielded from the crystal electric field effects by the outlying $5s$ and $5p$ electrons. The magnetic susceptibilities then have values which may be calculated as if the ions were in the ideal gaseous state.

The difference between these two extreme cases of spin-only magnetic susceptibility and the values obtained for f-electrons with full orbital and spin contributions is shown in Fig. 7. Also shown in the figure are the experimentally determined susceptibilities of the actinide ions in aqueous solution—a nearer approach than most solid compounds to the dilute environment implied by the theoretical calculations using the gaseous state. It is evident that the observed

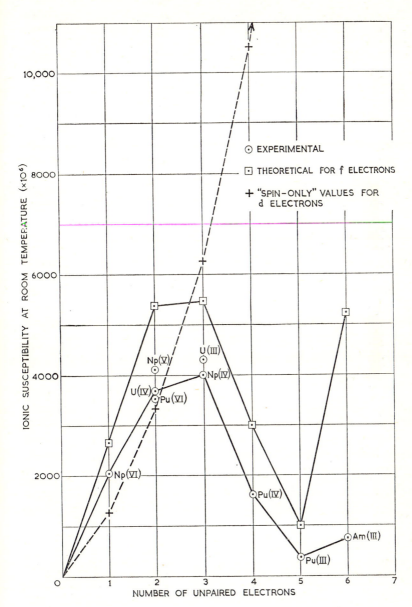

Fig. 7. The Magnetic Susceptibilities of Actinide Ions in Solution compared with the Extreme Theoretical Values

magnetic susceptibilities follow reasonably closely those predicted for $5f$ electron configurations; in ions containing less than three unpaired electrons there is insufficient difference between the predicted values for $5f$ and $6d$ electrons to make definite conclusions from the observed values. A much more complex theoretical treatment is required for these cases. Such a treatment in the case of the ions UO_2^{2+}, NpO_2^{2+} and PuO_2^{2+} has been made. The observed magnetic susceptibility of the plutonyl ion (PuO_2^{2+}) in the solid compound sodium plutonyl acetate has the *spin-only* value. Detailed calculations which took into account all the bonding characteristics and special properties of this type of oxygenated ion showed, however, that the observed value is entirely consistent with a $5f^2$ electron configuration.

In order to render comparison with theory more valid, magnetic susceptibility measurements on solid compounds are often made in a series with increasing dilution of the paramagnetic ion by an isomorphous diamagnetic ion and extrapolating the results to infinite dilution. This technique has been applied to some compounds of the actinides by making use of the isomorphous series of crystalline compounds mentioned above. Thus ThF_4 is diamagnetic and

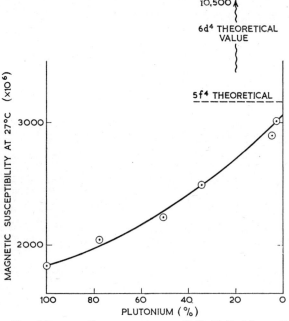

FIG. 8. THE MAGNETIC SUSCEPTIBILITIES OF PuF_4–ThF_4 MIXED CRYSTALS

contains no unpaired electrons, it may be used to dilute the iso-morphous PuF_4. The results of magnetic susceptibility measurements on the mixed crystals are illustrated in Fig. 8, which shows that although undiluted PuF_4 has a rather low susceptibility the extrapolation to infinite dilution agrees very well with the theoretical value for a $5f^4$ electron configuration.

The electron configurations which may be taken as having been most firmly established by magnetic susceptibility measurements are shown in Table 13. Trivalent uranium is included here, since although the experimental evidence from susceptibility measurements on powdered solid compounds was inconclusive, paramagnetic resonance work with dilute single crystals (e.g., UCl_3 in $LaCl_3$, and UF_3 in CaF_2) have shown characteristics very similar to those of trivalent neodymium in the same environment. Confirmatory paramagnetic resonance data have also been obtained on NpO_2^{2+}, PuO_2^{2+} and Cm^{3+}.

TABLE 13.—ELECTRON CONFIGURATIONS OF ACTINIDE
IONS FROM MAGNETIC SUSCEPTIBILITY MEASUREMENTS

No. of 5f Electrons	0	1	2	3	4	5	6	7
Uranium	UO_2^{2+}			U^{3+}				
Neptunium		NpO_2^{2+}	NpO_2^+					
Plutonium			PuO_2^{2+}		Pu^{4+}	Pu^{3+}		
Americium							Am^{3+}	
Curium								Cm^{3+}

The measurements on curium which indicated that the Cm^{3+} electron configuration was $5f^7$ are worthy of special mention since they were obtained with only a very few μg of curium trifluoride. The sensitivity of many magnetic susceptibility balances is such that a sample size of the order of grams is often required. One of the authors found it necessary to develop a susceptibility balance which could operate on samples down to one mg for work on plutonium; the sensitivity of the method was increased still further by Crane, Wallman and Cunningham at the University of California for their investigations on curium. They found that microgram samples of curium trifluoride diluted in solid solution in diamagnetic lanthanum trifluoride had a molar magnetic susceptibility of 26 500 c.g.s. units at room temperature, compared with 26 000 for the $5f^7$ theoretical value. One of the difficulties of physical measurements of this type on curium is the accurate measurement of the temperature of the samples since there is considerable self-heating due to the intense alpha-particle emission.

Evidence from magnetic susceptibility measurements supports the possibility of electrons occupying $6d$ energy levels, rather than $5f$, in compounds of thorium in its trivalent state and in quadrivalent uranium.

EMISSION SPECTRA

The methods mentioned in the preceding sections have given data on the electronic configurations associated with the ions of the actinides. Emission spectroscopy has given evidence not only on ionised species but also on the neutral atoms.

The emission spectrum of un-ionised actinium has been observed recently by Meggers, Fred and Tompkins at the U.S. National Bureau of Standards and has been interpreted in terms of a ground state of $6d7s^2$ for the three unpaired electrons. The singly and doubly ionised actinium spectra indicated corresponding ground states of $7s^2$ and $7s$.

Other workers have shown that the ground state of un-ionised thorium is $6d^27s^2$, but that a $5f^1$ ground state occurs in trivalent thorium. The $5f$ levels become much more stable in uranium and the un-ionised spectrum then corresponds to a ground state of $5f^36d7s^2$. Un-ionised americium has been shown to have the ground state $5f^77s^2$.

Thus the emission spectra confirm the increasing stability of the $5f$ electron level with movement up the actinide series.

THE TRANSCURIUM ELEMENTS

At the time when Seaborg put forward the arguments for the actinide concept in the open literature (1949) only the elements up to curium had been discovered. Using this concept, it was possible to predict the extensive similarities which should exist between the higher actinides and the lanthanides on the basis that both series of elements contained unpaired electrons in f energy levels. If the concept was valid, then the greater part of the solution chemistry of the transcurium elements should be predictable from a knowledge of the chemistry of the lanthanides. The latter elements were found to elute at different rates from each other after adsorption on cation-exchange resins (e.g. Dowex 50), and by means of equipment which fractionated the eluting solution into successive drops coming from the resin column, the elements could be separated. Application of the same technique to the actinides enabled Thompson, Ghiorso and Seaborg to predict which fraction of eluting solution might contain each particular element. This prediction, together with the relative ease and rapidity by which ion-exchange separations could be performed on a small scale, meant that there was a reasonable chance of observing quite short-lived isotopes of the higher elements.

The first transcurium element to be discovered was berkelium; the particular isotope produced (Bk^{243} by helium ion bombardment of Am^{241}) had a half-life of only 4·5 h; nevertheless, it was successfully separated by the ion-exchange method and identified by alpha-pulse analysis. Similarly each successively higher actinide element has been separated and identified using the ion-exchange technique, culminating in the discovery of mendelevium (Mv^{256}) with an electron-capture half-life of 30 min., and 254 isotope of element 102 with an alpha decay half-life of 3 seconds.

The production of elements of higher atomic number than 102 will obviously be extremely difficult and the use of cyclotrons and linear accelerators to give heavy ions of very high energy for bombardment of lower actinide element targets will be the most probable route for their syntheses. Multiple neutron capture is not a promising route, since some of the intermediate isotopes in the fermium region have extremely short half-lives and would not exist long enough under irradiation to further the build-up of higher elements. The irradiation resulting from exposure of the lower actinides to very high neutron fluxes for extremely short times in thermonuclear bomb explosions has overcome this difficulty and has produced identifiable amounts of einsteinium and fermium in the fallout debris, but this can scarcely be considered a promising route for further progress!

CHEMISTRY OF THE ACTINIDE ELEMENTS

Excellent reference books are available which give fully detailed accounts of the chemistry of these elements and a repetition of this data is beyond the scope of the present book. Several rather isolated topics will be discussed, however, in the cases where the compounds or solutions concerned are of special importance in the atomic energy programme.

THE METALS

The metallic elements from actinium to curium have been prepared, but extensive data are not available on actinium, protactinium or curium. The melting points of the metals are:

Ac	Th	Pa	U	Np	Pu	Am
1050	1750	?	1132	640	640	?°C

Thorium, uranium and plutonium are of vital concern to the atomic energy industry and, of these three, thorium is the most difficult to prepare in a pure state, owing to its high melting point. Moreover, thorium is highly reactive in the molten state and is difficult to contain. Either the oxide or the tetrahalides may be reduced by calcium. Normally the temperature attained in carrying out this reaction is insufficient to melt the thorium and the latter is then produced in a finely divided and pyrophoric form. Investigations at Ames Laboratory, U.S.A., have shown that it is advantageous to add zinc chloride to the original reactants (ThF$_4$ and metallic calcium)—a lower melting zinc-thorium alloy is formed which can be recovered as ingots.

Uranium metal may similarly be prepared by the reduction of the oxide or a tetrahalide with an electropositive metal. It is possible to produce the metal in ingot form by performing the reduction under such conditions that the whole reaction mass reaches the molten state. This is a less formidable task than for thorium, since the melting point is lower. If the starting material is uranium tetrafluoride then the use of magnesium or calcium results in the formation of the corresponding fluorides which have reasonably low melting points (e.g. MgF$_2$, 1263°C). For this reason, the fluoride system is usually preferred to that of reduction of uranium oxides since the latter form refractory oxides with the electropositive metals and these hinder the aggregation of the molten uranium. Of the two most favoured reducing metals, calcium and magnesium, the former has more desirable thermochemistry:

$$UF_4 + 2\ Ca \rightarrow U + 2\ CaF_2 \qquad \Delta F_{298} = -135\ kcal$$
$$UF_4 + 2\ Mg \rightarrow U + 2\ MgF_2 \qquad \Delta F_{298} = -79\ kcal$$

However, magnesium is easily obtained in pure form and does not react with air during preliminary handling, and on these grounds it is often considered to be the most suitable reductant.

Plutonium metal was first produced at the University of Chicago in 1943. The work was done on a very small scale and the metal was obtained in the form of globules weighing less than 50 μg. Suitable techniques were developed for the measurement of density and melting point and for the observation of X-ray diffraction patterns using these minute samples.

At the present time plutonium is produced on an industrial scale in the U.S.A., the United Kingdom, the U.S.S.R. and France (See Plates II and III). Details of the methods have not been published, but in principle the reduction of the oxide or halides with electropositive metals is possible as in the case of thorium and uranium.

The number of possible reacting combinations is greater with plutonium owing to the marked stability of the trivalent halides.

Uranium and plutonium have complex structures which do not correspond to any of the lanthanide elements. Moreover, uranium exhibits three allotropic modifications between room temperature and the melting point, and plutonium can exist in the remarkably high number of six modifications below its melting point. The structure types and transition temperatures are shown in Table 14.

TABLE 14.—ALLOTROPIC FORMS OF URANIUM AND PLUTONIUM

Phase		Lattice Type	Temperature Range (°C)
Uranium:	Alpha	Orthorhombic	r.t.–668
	Beta	Tetragonal	668–774
	Gamma	Body-centred cubic	774–1132
Plutonium:*	Alpha	Monoclinic	r.t.–125
	Beta	?	125–220
	Gamma	Face-centred orthorhombic	220–320
	Delta	Face-centred cubic	320–450
	Delta-prime	Body-centred tetragonal	450–480
	Epsilon	Body-centred cubic	480–640

* Approximate values. There is considerable variation between the values quoted from U.S., U.S.S.R. and U.K. investigations.

This complexity of form contributes to the problems of designing reactor fuel elements for long-life—for instance thermal cycling, such as would occur in a nuclear reactor, of uranium through the alpha-beta transformation temperature has been shown to give severe distortion of a uranium metal bar (See Plate VIII). Thermal cycling of plutonium gives a similar distortion.

Apart from the large number of allotropic forms, plutonium also is unique in another respect. No other pure metal having isotropic crystal structures exhibits the *negative* coefficient of thermal expansion found for the delta and delta-prime phases. The negative coefficient has not yet been fully explained theoretically.

Both uranium and plutonium are extremely reactive metals. Their silvery lustre is rapidly lost on exposure to air; uranium acquires an adherent black oxide film whereas plutonium gives a loose deposit of the dioxide. Relatively massive lumps of plutonium are warm to the touch owing to the intense alpha particle emission within the metal. Plutonium metal can never be handled on a really large scale in the same way as natural uranium, since nuclear criticality considerations set an upper limit to the amount which may be handled at any one time. Assessing the upper limit in a particular set of circumstances is a complex task, since criticality depends not only on the metal itself but also on the surrounding materials, for instance special consideration must be given to the presence of good neutron reflectors such as water or graphite— often used as crucibles for metallurgical work.

THE FLUORIDES

The actinide fluorides have considerable technological import-ance: UF_6 is used for the separation of U^{235} from natural uranium, UF_4 is used as the reactant in one of the principal methods of uranium metal production and differences in stability and volatility between uranium and plutonium fluorides are utilised in the fluoride processing of reactor fuel rods (Chapter 5).

The following fluorides are known to exist:

AcF_3		UF_3	NpF_3	PuF_3	AmF_3	CmF_3
	ThF_4 PaF_4	UF_4	NpF_4	PuF_4	AmF_4	CmF_4

$$(PaF_5) ? U_4F_{17}$$
$$U_2F_9$$
$$UF_5$$
$$UF_6 \quad NpF_6 \quad PuF_6$$

The three hexafluorides are solid at room temperature, but sublime at slightly higher temperatures. Uranium hexafluoride is the most stable of the three and plutonium hexafluoride is the least stable. This is reflected also in the relative ease of preparation of UF_6 compared with PuF_6, thus the reaction

$$XF_4 + F_2 \rightarrow XF_6$$

proceeds at a reasonable rate at 300 to 400°C for uranium, but for plutonium the reaction is slow below a temperature of 750°C. Both of the hexafluorides may be formed by the interesting reaction of the corresponding tetrafluoride with oxygen:

$$2 XF_4 + O_2 \rightarrow XF_6 + XO_2F_2$$

The oxyfluoride remains in the fluorination apparatus, whereas the hexafluoride distils out. This reaction proceeds at about 800°C and, both in this case and in the reaction with fluorine at the same temperature, the plutonium hexafluoride is extremely reactive towards the constructional materials. Under very dry conditions and in the absence of hydrogen fluoride, both UF_6 and PuF_6 may be handled in glass apparatus at lower temperatures. Hydrolysis of PuF_6 by water or moist air at room temperature is violent and is accompanied by flashes of light. Even though, from a compatibility viewpoint, the PuF_6 may be stored for long periods in dry containers, a free-flowing pink solid appears after a few days together with a rise in the system pressure. This is due to decomposition of PuF_6 to PuF_4 and fluorine by the alpha particles emitted as a result of radioactive decay. The rate of disappearance of PuF_6 from this cause is about 1·5 per cent per day.

The tetrafluorides vary considerably in their ease of preparation; thus the tetravalent salt is the only known fluoride of thorium and is easily prepared by reaction of the oxide with gaseous hydrogen fluoride, whereas americium tetrafluoride requires the use of fluorine. Uranium and plutonium tetrafluorides are usually prepared by gaseous hydrofluorination of the corresponding oxide in a reactive state with a high surface area (obtained by low-temperature calcination of thermally unstable compounds). Hydrogen fluoride obtained from steel cylinders usually has traces of hydrogen and sulphur dioxide as impurities; these are sufficient to produce plutonium trifluoride by reaction such as:

$$PuO_2 + 3HF + \tfrac{1}{2}H_2 \rightarrow PuF_3 + 2H_2O$$

In order to produce the tetrafluoride it is necessary to add oxygen to the hydrogen fluoride stream. The reaction between hydrogen fluoride and the dioxides of uranium and plutonium at temperatures less than 100°C follows the course:

$$XO_2 + 4HF \rightarrow XF_4 . 2H_2O$$

The product has a crystal structure very similar to the original cubic dioxide and it is only above 200°C that the anhydrous monoclinic tetrafluoride is obtained.

Precipitation from aqueous solution by aqueous hydrofluoric acid gives hydrates with a different crystal structure and which have the formulæ UF_4, 2·5 H_2O and PuF_4, 2·5 H_2O. The enhanced stability of the trivalent state for plutonium is illustrated by the fact that attempts to dehydrate this hydrate in a vacuum lead to the reaction expressed approximately by:

$$4 PuF_4 + 2H_2O \rightarrow 4 PuF_3 + 4HF + O_2$$

Uranium trifluoride is more difficult to prepare; reaction of the tetrafluoride with hydrogen must be performed at about 1000°C under conditions which are carefully controlled to avoid hydrolysis by traces of moisture. A more convenient preparation is by the high temperature reduction of the tetrafluoride with metallic uranium.

The fluoride systems offer an interesting example of the care which must be exercised in applying the actinide concept too rigorously to the early members of the series. For instance, there is no ThF_4, $2\cdot5$ H_2O to correspond to the uranium and plutonium compounds, $ThF_4.H_2O$ is not isomorphous with $UF_4.H_2O$, and $ThOF_2$ has much greater stability than the uranium oxyfluorides. The thorium fluoride system resembles that of zirconium more closely than those of the higher actinides. A reactor design which utilises molten fluorides at high temperatures is discussed in Chapter 9.

THE OXIDES

The actinide oxides are technologically important as intermediates in the preparation of other compounds and as the probable basis of future types of long-lived fuel elements for nuclear reactors. Definite oxide structures which have been elucidated correspond to the following formulæ:

		(UO)		(PuO)		
Ac_2O_3				Pu_2O_3	Am_2O_3	Cm_2O_3
	ThO_2	PaO_2	UO_2	NpO_2	PuO_2	AmO_2	CmO_2
		Pa_2O_5	U_4O_9				
			U_3O_8	Np_3O_8			
			UO_3				

The most noticeable feature is the exceptional stability of the fluorite-type lattice of the dioxides. With the exception of actinium, which cannot have a valency greater than three, all of the actinides which are available for quantitative chemical investigation show this type of dioxide.

The uranium-oxygen system is the most complex amongst the actinides owing to the large number of possible stable valency states. Also, there are substantial deviations from stoichiometry of these oxides; for instance the fluorite crystal structure of UO_2 is retained over the concentration range $UO_{1\cdot75}$ to $UO_{2\cdot25}$ under certain conditions, and uranium dioxide which has been exposed to the air at room temperature has an oxygen excess which varies according to the surface area of the sample. Gravimetric analyses of uranium are commonly based on ignition in air to U_3O_8; the constancy of weight achieved is fortuitous since the composition has been found to vary with the ambient oxygen pressure.

Plutonium dioxide results from the ignition of a wide variety of plutonium compounds in air, deviations from stoichiometry above the composition $PuO_{2\cdot00}$ have been observed, but all attempts to produce a definite higher oxide have failed.

The rather complex system of hydrates formed by uranium

trioxide in high temperature water is mentioned in Chapter 7. The dioxides do not form stable hydrates under these conditions.

DISPROPORTIONATION REACTIONS IN SOLUTION

Just as the multiplicity of valency states renders the chemistry of the solid compounds such as the oxides complex, so we have also considerable complexity in the solution chemistry of plutonium. The intermediate oxidation states of plutonium in solution can undergo self-oxidation and reduction reactions. The electrode potentials show that the couples between the various oxidation states are all relatively close and they may be varied by changing the pH and anion concentration. In a single solution it is possible to calculate the equilibrium proportions of the various oxidation states which will be taken up; for instance in perchloric acid the relative proportions are:

Pu Oxidation State	III	IV	V	VI
0·1 M HClO$_4$	19	8	13	60 per cent
1·0 M HClO$_4$	30	60	0	10 per cent

Evidently as many as four oxidation states may exist simultaneously in solution. A solution which was originally prepared as Pu(IV) will disproportionate until the solution reaches equilibrium. The rate-controlling step in this disproportionation is

$$2 Pu^{4+} + 2H_2O \rightleftharpoons Pu^{3+} + PuO_2^+ + 4H^+$$

and this is followed by the rapid reaction

$$Pu^{4+} + PuO_2^+ \rightleftharpoons Pu^{3+} + PuO_2^{2+}$$

The situation is further complicated by the fact that Pu(V) also undergoes disproportionation

$$2 PuO_2^+ + 4H^+ \rightarrow Pu^{4+} + PuO_2^{2+} + 2H_2O$$

In order to maintain a particular oxidation state (such as is required, for instance, in the separation of plutonium from uranium) it is necessary to add an oxidising or a reducing agent to the system to overcome these effects of disproportionation.

The other quinquevalent actinide ions, U(V), Np(V) and Am(V) also appear to undergo similar disproportionation reactions. The overall reaction for AmO$_2^+$ seems to be

$$3 AmO_2^+ + 4H^+ \rightarrow 2AmO_2^{2+} + Am^{3+} + 2H_2O$$

The rate-controlling step is presumably the disproportionation of two Am(V) ions into Am(IV) and Am(VI), but as the quadrivalent ion is unstable in aqueous solution it becomes immediately reduced to Am(III) possibly by reducing agents produced in the solution by alpha radiation or possibly by the reaction

$$Am^{4+} + AmO_2^+ \rightarrow AmO_2^{2+} + Am^{3+}$$

The disturbance of the equilibria between the various valency states due to radiation decomposition of the water becomes quite pronounced with americium. Auto-reduction of PuO_2^{2+} due to alpha radiolysis of the solution is detectable, for instance it is 0·2 per cent per day in 0·5 M HCl. The alpha-disintegration rate of Am^{241} is rather more than fifty times that of Pu^{239} and the rate of auto-reduction of AmO_2^{2+} is correspondingly higher; the rate depends very much on the composition of the solution, however. When Cm^{242} is being handled, the radiation decomposition of the water becomes so great that it has been found to be impossible to oxidise americium to AmO_2^{2+} in the presence of $1·5 \times 10^{-4}$ M curium.

It is hoped that this brief account of the chemistry of the heaviest elements has indicated some of their novel characteristics and shown why they are usually classed together as a series analogous to the lanthanides. If the analogy between the lanthanides and the actinides holds, only one other element—number 103—remains to be isolated to complete the series.

FURTHER READING

Katz and Rabinowitch. *The Chemistry of Uranium.* National Nuclear Energy Series, VIII-5. McGraw-Hill Book Co. Inc., 1951.

Katz and Seaborg. *The Chemistry of the Actinide Elements.* Methuen and Co. Ltd., 1957.

United Nations. Proceedings of the First International Conference on the Peaceful Uses of Atomic Energy, Geneva, 1955. Vol. 7, *Nuclear Chemistry and the Effects of Irradiation.*

CHAPTER 5

SEPARATION PROCESSES

Apart from the obvious separation processes involved in the extraction and production of the main constituents of nuclear reactors, there are two types of separation which are of vital importance to the atomic energy programme. The first of these is the mutual separation of plutonium, uranium, and fission products from irradiated-uranium fuel elements and the second is the separation of uranium from irradiated thorium. Historically the treatment of irradiated uranium has been the most important and will be discussed first.

TYPES OF PROCESS FOR IRRADIATED URANIUM

Up to the present time, the vast majority of nuclear reactor fuel elements have been made of metallic uranium. This has serious limitations, however, since the time a fuel element may remain in the reactor is then limited by two factors: the build-up of neutron poisons and physical damage to the metal (see Chapter 3).

In order to obtain a reasonable utilisation of the U^{235} constituent of the fuel, from time to time it becomes necessary to remove the irradiated fuel elements from the reactor, separate off the fission-product neutron poisons, reconstitute the fuel rods and continue the irradiation. It may be necessary to repeat this whole cycle several times on one particular batch of uranium. Moreover, the U^{238} part of the fuel elements will have built up plutonium by neutron capture processes and it is necessary to recover this either for the weapons programme or as a valuable alternative fuel for future power reactors.

Many processes have been proposed for the separation of plutonium from metallic uranium and the decontamination of both of these from fission products; they may, however, be classified into a small number of types:

> Precipitation
> Ion Exchange
> Solvent Extraction
> Volatilisation
> High-temperature Pyrometallurgy

The first three methods depend upon an initial step in which the uranium must be converted to an aqueous solution of one of its salts. For protection within the reactor the fuel elements have a metallic sheath and, according to the particular metal employed, there is a variety of mechanical or chemical methods used to remove this before the separation process can begin. In early types of fuel element the can was often made of aluminium and this could be removed by dissolution in aqueous caustic soda. As fuel element design has progressed more exotic materials, such as stainless steel and niobium, have been used for the cans, thereby producing much more difficult problems for this first stage.

AQUEOUS PROCESSES

The dissolution of the decanned uranium metal, preparatory to the separation procedure, may be accomplished by heating with an excess of strong nitric acid in stainless steel equipment. In the early work on this method, large quantities of oxides of nitrogen were evolved and these were difficult to deal with, particularly since they carried off fission product xenon from the dissolver.

The stoichiometry of the reaction depends upon prevailing conditions in the dissolver; with $11 \cdot 7$ M nitric acid the reaction may be represented approximately by the equation:

$$U + 4 \cdot 5\ HNO_3 \rightarrow UO_2(NO_3)_2 + 1 \cdot 57\ NO + 0 \cdot 84\ NO_2$$
$$+ 0 \cdot 0005\ N_2O + 0 \cdot 043 N_2 + 2 \cdot 25\ H_2O$$

Fumeless dissolving was applied later, however—an extension of a method originally used commercially in the production of silver nitrate from metallic silver. The principle of this method is that a packed reflux column is fitted above the dissolving vessel; oxygen fed in at the bottom of the column converts the oxides of nitrogen back to nitric acid and thereby makes for a much cleaner process, with the additional advantage that nitric acid consumption is cut by a factor of at least two:

$$U + 2HNO_3 + \tfrac{3}{2}O_2 \rightarrow UO_2(NO_3)_2 + H_2O$$

The world's first large-scale plant for plutonium separation was built at Hanford, U.S.A., in 1945; an aqueous precipitation process was employed since at that time it was the only well-established method. It had the disadvantage that fission product activity was not separated from the residual uranium which consequently could not be reused. Solvent extraction methods which did not have this limitation were devised later both in the U.S.A. and at Chalk River, in Canada.

(i) Precipitation

Several types of precipitation reactions have been used during laboratory investigations of the very heavy elements and one of the standard methods for the estimation of plutonium involves a co-precipitation of the trivalent ion with lanthanum fluoride, a separation from uranium being achieved thereby. The particular precipitation process which attained production scale status in the U.S.A. is known as the Bismuth Phosphate Process.

Uranyl nitrate obtained by dissolution of irradiated uranium metal, and containing plutonium and the longer-lived fission products, is treated with sodium nitrite to ensure that all the plutonium is in the quadrivalent state and sulphate ions are added to complex the uranium. Bismuth and phosphate ions are then separately added to the solution to precipitate bismuth phosphate which carries the plutonium, leaving most of the uranium and some of the fission-product activity in solution. Further decontamination is achieved by dissolving the precipitate in nitric acid and twice repeating the cycle, which then includes a further bismuth phosphate precipitation with the plutonium held in the soluble hexavalent state by the addition of, for instance, permanganate or dichromate ions.

Enhanced decontamination from radioactive fission-product zirconium is obtained by the addition of inactive zirconium in the precipitation steps when the plutonium is hexavalent (the zirconium follows the bismuth phosphate) and by the addition of fluosilicate ions to retain the zirconium in solution during the co-precipitation of bismuth phosphate with reduced plutonium. One cycle of the precipitation process is illustrated in Fig. 9.

The process ends with a further cycle using lanthanum fluoride precipitation as a carrier for the plutonium and this precipitate is finally converted into a mixture of plutonium and lanthanum nitrate solutions by metathesis of the fluorides with aqueous potassium hydroxide to form the nitric acid soluble heavy metal hydroxides. A final separation of plutonium from lanthanum is achieved by precipitation of the former as a peroxide, lanthanum remaining in solution.

The decontamination of plutonium from fission product activity is about 10^7, and overall recovery of plutonium from very dilute feed solutions has been quoted as better than 95 per cent for this process. Solvent extraction methods can do better than this, however, and are more amenable to large scale continuous operation.

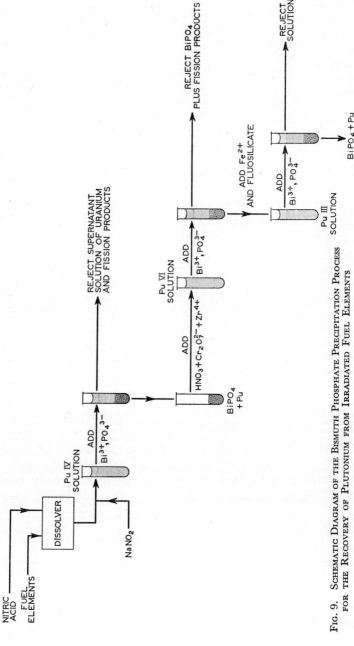

Fig. 9.   Schematic Diagram of the Bismuth Phosphate Precipitation Process for the Recovery of Plutonium from Irradiated Fuel Elements

(ii) Ion Exchange

Ion exchange processes, although of great value for laboratory separations, have not been developed for large-scale use on highly radioactive solutions owing to the sensitivity of the organic resins to breakdown by radiation and plugging of the columns by gases evolved in the radiation decomposition of water. Consequently they will not be discussed in detail in this chapter.

(iii) Solvent Extraction

Although a joint Canadian and United Kingdom team had begun preliminary investigations on solvent extraction and other methods in Montreal during 1944, it was not until 1946 that a firm requirement for a British separation plant was formulated. Further investigations at Chalk River then showed that the most promising method for separating both plutonium and uranium from fission product activity, and which could most readily be scaled up from laboratory work to factory process, was that of partition of the species concerned between an aqueous phase and an organic liquid phase. This was a process which was well known in organic chemistry, but which up to that time had not received very much attention in the inorganic field.

The organic solvent which was finally chosen for use at Windscale was dibutyl carbitol (butex). Other organic solvents upon which considerable development work has been done are methylisobutyl ketone (hexone), tri-n-butyl phosphate (TBP) and triglycol dichloride (trigly). One of the reasons for choosing butex in 1947 was that it had been found to be relatively more stable than some of the other solvents towards fairly strong nitric acid solutions. Moreover, the desired partition characteristics could be obtained from a 3 N nitric acid solution: solvents like hexone and trigly required high concentrations of aluminium nitrate in the aqueous phase to act as a *salting-out* agent for the plutonium, this in turn led to difficulties in concentrating the plant waste solutions containing the fission products.

The solvent extraction processes depend upon the variable valency of plutonium. The hexavalent state (PuO_2^{++}) is soluble in the organic solvents, together with uranium in the uranyl form. By a suitable choice of reducing agent plutonium can be converted to the trivalent state without reducing the uranium. Trivalent plutonium is soluble in dilute aqueous acids, but almost insoluble in the organic phase. Consequently the final form into which the Windscale process developed included this plutonium reduction step and the whole process is illustrated diagrammatically in Fig. 10.

Countercurrent extraction columns are used so that the whole process may be run on a continuous basis.

After a suitable storage time, irradiated fuel elements are fed into the dissolver and are reacted with nitric acid. The resultant solution is then diluted and fed into the first extraction column. The conditions in the dissolver having produced both uranium and plutonium in an oxidised state, both of these elements then extract into the organic solvent which is fed into the bottom of the extractor. The vast majority of the fission-product activity remains in the aqueous phase which is transferred to an evaporator and then to storage tanks.

Dissolved nitric acid in the organic stream is neutralised and the plutonium is then reduced to the trivalent state by ferrous sulphamate. Contact with aqueous $0.27N$ HNO_3 containing $8M$ NH_4NO_3 then separates the plutonium from the uranium, which remains in the organic phase. The latter is stripped into dilute acid (Extractor 3), which is then concentrated by evaporation and transferred to a separate purification system where the last traces of radioactivity are removed to enable the recovered uranium to be handled in the same way as unirradiated material for reconversion to fuel elements.

The aqueous plutonium-bearing stream is oxidised with sodium dichromate and again extracted into the organic phase (Extractor 4), giving a further decontamination from fission-product activity. The final extraction column recovers plutonium in dilute acid, which is then concentrated by evaporation before proceeding to the plutonium purification plant. From here onwards, the gamma radiation from the plutonium solution is so low that expensive concrete and lead shielding can be dispensed with, further handling being performed in glove boxes.

The production timetable which was laid down in 1947 for the Windscale plant was so tight that it became necessary to begin the general design of the plutonium extraction plant costing several million pounds on the basis of the laboratory experiments at Chalk River involving only a few mg of plutonium. Further intensive efforts with larger quantities of materials, particularly with higher fission-product activities, subsequently played a vital part in enlarging the designers' knowledge of the details of the operations involved.

At about the same time as the preliminary work for Windscale was proceeding at Chalk River, a very similar process was being developed in the U.S.A. based upon hexone. The latter has been termed the *Redox* process and the main differences from the butex

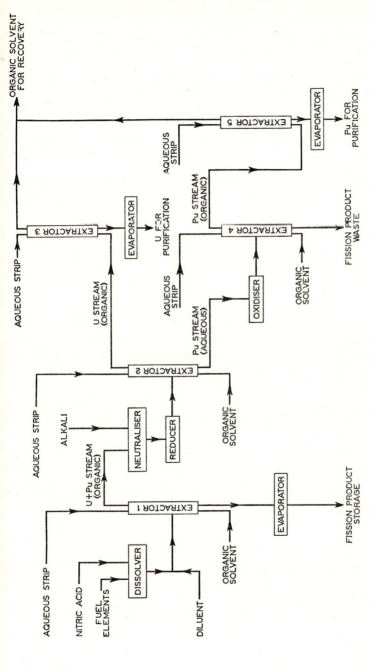

FIG. 10. SCHEMATIC DIAGRAM OF THE WINDSCALE PRIMARY SEPARATION PLANT FOR THE RECOVERY OF URANIUM AND PLUTONIUM FROM IRRADIATED FUEL ELEMENTS

process arise from the decreased stability of hexone towards nitric acid. Only very low acidities may be tolerated, and this could give rise to partial hydrolysis of plutonium in the tetravalent state; hence it is necessary to oxidise the plutonium to plutonyl before extraction. Also as a result of the low acidity, an inorganic nitrate must be added to the aqueous phase to obtain the desired extraction coefficient. Aluminium nitrate is the usual material used for this salting-out.

Typical partition coefficients for extraction into hexone from $1\cdot0$ M Al $(NO_3)_3$ solution adjusted to the desired pH by additions of nitric acid or caustic soda are quoted below from a paper by R. E. Tomlinson at the first Geneva Conference on the Peaceful Uses of Atomic Energy:

pH	Partition Coefficient		
(Aqueous Phase)	U VI	Pu VI	Pu IV
0·0	2·1	3·2	3·0
0·5	1·6	2·9	0·85
1·5	1·0	2·5	0·09
2·5	0·6	2·0	0·01

It is evident that while U VI and Pu VI are extractable at relatively high pH, the efficiency of extraction of Pu IV falls very rapidly with increasing pH, due to hydrolysis to an inextractable form.

The Redox process then consists of oxidising plutonium in the dissolver solution, neutralising and adding aluminium nitrate, and extracting the plutonium and uranium into the hexone phase in the first column. By subsequently reducing the plutonium to the trivalent state it is stripped from the hexone into an aqueous phase in a second column; this aqueous phase again must contain a high concentration of aluminium nitrate to retain the uranium in the hexone. A third column is used to strip the uranium into dilute nitric acid. After achieving this initial separation it is necessary to put both uranium and plutonium through further cycles of extraction in order to obtain further purification. One of the major disadvantages of the Redox process is that the fission-product waste solutions contain large quantities of inorganic nitrate and the disposal problems are thereby intensified considerably.

More recent work in both Britain and the U.S.A. has shown that tri-n-butyl phosphate has advantages over both butex and hexone as an extractant. Decontamination factors from fission-product activity are much higher at each stage, and this can lead to a considerable reduction in the overall size of the extraction plant and hence to a smaller processing cost. Plutonium has been shown to extract into the organic phase as the complexes $Pu(NO_3)_4 \cdot 2TBP$ and $PuO_2(NO_3)_2 \cdot 2TBP$ from aqueous nitric acid solutions of the tetravalent and hexavalent states. An easy separation may be achieved between tetravalent plutonium and hexavalent uranium.

(iv) Limitations of the Solvent Extraction Process

Although the solvent extraction processes have been vital to the furtherance of the atomic energy industry, there are several problems which are inherent in their use and there are limitations which may ultimately mean their replacement by other types of process.

Hydrolysis has been mentioned above as one of the important factors affecting the design of the Redox process. It can affect the TBP process also, but in a different manner. In that case it is the organic material which undergoes slow hydrolysis with the production of dibutyl phosphate. The latter forms much more stable complexes with many of the metal ions present in the aqueous phase than the complexes formed with the original tributyl phosphate, hence it becomes increasingly difficult to strip these metals back into an aqueous phase as the concentration of DBP builds up. Much of the early work on TBP extraction appeared very confusing until the marked effect of slight traces of DBP had been elucidated.

Much effort has been required to determine the behaviour of fission-product ruthenium in extraction processes involving all of the above-mentioned solvents. The dissolution of the metallic uranium fuel element appears to produce at least three distinct forms of ruthenium in the resultant solution: a solvent-soluble form, a less soluble form and a relatively inextractable form. Moreover, considerable lengths of time are required to establish an equilibrium between these forms. After the first extraction column, ruthenium is one of the most important remaining contaminants of the uranium and the plutonium, and one which has been difficult to deal with because of its complex chemistry. Investigations have shown that the solvent-soluble form is a complex of nitrosyl ruthenium:

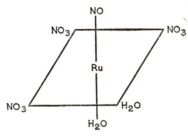

The dinitrato complex is the less solvent-soluble form while the mononitrato complex is almost insoluble. The situation is rendered even more complicated by the production of nitro complexes due to the presence of traces of nitrous acid during the dissolving stage. It can be seen that a large effort has been required in order to clarify the relevant chemistry of ruthenium, but the situation is now reasonably satisfactory and this element is no longer a major problem.

Trouble has arisen from time to time due to certain nitric-acid insoluble impurities in the uranium metal. Unless the purity of the metal is rigorously controlled, quite small amounts of such materials as silica will enter the solvent extraction plant. If these occur as discreet particles of very small diameter, in the region of one micron, they will tend to congregate at the interface between the aqueous and the organic phases in the separation columns and will decrease decontamination factors by adsorbing ions of some of the important fission products, such as zirconium.

It is well known that most organic materials are readily degraded by the action of ionising radiations. The effects of the fission product beta and gamma activity on the organic phase, particularly in the first extraction column of the separation processes, has required considerable investigation. It has been found, for instance, that these radiations will speed up the production of the dibutyl and monobutyl derivatives of tributyl phosphate. The effect of the build-up of dibutyl phosphate is then to cause increased retention of extractable complexes in the organic phase and the monobutyl phosphate forms insoluble complexes with uranium and plutonium. As little as 0·01 per cent of dibutyl phosphate has been shown to decrease the ease of stripping fission-product zirconium from an organic phase containing 20 per cent TBP diluted in kerosene (the latter added to obtain low viscosity). Provision has to be made for removal of the DBP and MBP from TBP which is to be recycled, by washing with sodium carbonate solution. Estimates have been made that processing of highly irradiated uranium after only ten days cooling time, would not irradiate the solvent at a rate sufficient to affect its extraction properties. It would seem, then, that solvent extraction methods will find wide application in the power reactor programme.

However, the effects of radiation on the organic phase are only negligible in very dilute systems and this means a large-volume plant. Moreover, the plant must be surrounded by very costly concrete shielding. In order to reduce the required shielding to reasonable proportions it is necessary to allow considerable time to elapse between removal of the uranium elements from a reactor and the processing cycle; this may be three months or more before the shorter-lived fission products have decayed to a sufficiently low level. In its simplest form the fuel cycle may look like this:

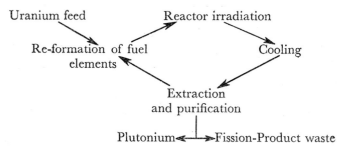

The relative amount of valuable fissile material essential to the process but not actually in use in the reactor is quite high and can add a significant amount to the cost of power produced from the reactor system. Hence, any device for increasing the recycling rate of the fissile material will be beneficial economically. Non-aqueous processes offer the possibility of very considerable reduction in the cooling time required, since they are able to operate at much higher radiation levels. Moreover, the overall volume of the plant required is likely to be much less than for solvent extraction. Some of the proposed processes are described in the following part of this chapter.

There is a further important point concerning the allowable cooling time for highly irradiated fuel. During irradiation, U^{237} has been formed in the uranium by the process of neutron capture in U^{235}. This isotope emits beta and gamma radiation and, of course, follows the main uranium stream through the extraction and purification processes. Since the half-life of U^{237} is 6·7 d, this implies that considerable activity will be associated with the purified uranium if the cooling time has been appreciably shorter than 100 d. Very short cooling times will mean that refabrication of the metallic fuel elements must be done entirely by remote control—a difficult and extremely expensive proposition. It is not clear at present which is likely to prove more acceptable economically: short cooling time with remote refabrication or long cooling time with direct refabrication.

VOLATILISATION PROCESSES

The disadvantage of the early bismuth phosphate precipitation as a separation method was that although plutonium was recovered the uranium was left with the highly radioactive fission products. Considerable work has been done on a volatilisation process involving fluoride systems and this suffers from the opposite disadvantage that uranium is easily separated, whereas the plutonium is more difficult to remove from the fission products. In an enriched uranium fuel system producing very little plutonium this would not matter; it is a serious problem with natural uranium fuels however. Nevertheless, the problem appears to be capable of solution and the fluoride volatilisation process has been taken to the pilot plant stage in the U.S.A.

The fluoride volatilisation process depends upon the high vapour pressure of uranium hexafluoride at relatively low temperatures (b.p. 56·5°C) and upon the fact that most of the fission-product fluorides are less volatile than this. Thus strontium, caesium, cerium and the lanthanide fluorides may be considered as non-volatile, whereas tellurium, ruthenium and niobium fluorides have an appreciable vapour pressure at the uranium distillation temperature. These elements may be removed subsequently by fractional distillation.

One of the most difficult problems of the fluoride volatilisation process has been the initial fluorination stage. Direct fluorination of irradiated metallic uranium metal with gaseous fluorine is not favoured since there is a high heat of reaction and it is difficult to remove this from the reaction zone rapidly enough for control of the reaction rate. Slower rates of reaction and an effective means of conducting heat away have been achieved by two suggested methods involving liquid phase fluorinations. The first method involves the dissolution of uranium metal in a molten NaF-ZrF_4 eutectic in the presence of excess hydrogen fluoride. The solubility in this system is appreciable and in fact this particular fluoride mixture has been proposed as the fuel system for an advanced type of nuclear reactor (see Chapter 9). If fluorine is now bubbled through the molten solution, the uranium may be converted to the volatile hexafluoride and distilled off, leaving most of the fission product activity in the fused salt.

The second method has received rather more attention and involves the dissolution of uranium in liquid bromine trifluoride. It has been observed that the metal must always be covered with the fluorinating liquid, otherwise the reaction may start to run away due to insufficiently rapid removal of the reaction heat. Both this reaction and the addition of fluorine to the molten salt mixture in the previous method give rise to extremely corrosive environments, but it has been possible to carry out the laboratory scale work in nickel and monel with the expectation of reasonably long equipment life. Bromine catalyses the reaction of uranium with bromine trifluoride, and so do compounds of the Group V elements. The latter fact arises from the ionic dissociation of bromine trifluoride:

$$2\,BrF_3 \rightleftharpoons BrF_2^+ + BrF_4^-$$

By analogy with the effect of hydrion on metal it seems reasonable that the effective fluorinating agent in bromine trifluoride attack is the BrF_2^+ ion. Hence the most facile attack is obtained by promoting the formation of this ion. This may be achieved by the addition of Group V elements, e.g.:

$$AsF_3 + BrF_3 \rightarrow BrF_2^+ + AsF_4^-$$

The fluorination reaction to form uranium hexafluoride may be represented by the following scheme:

$$2BrF_3 \rightleftharpoons BrF_2^+ + BrF_4^-$$
$$U + BrF_2^+ \rightarrow UF_4 + Br_2$$
$$UF_4 + BrF_4^- \rightarrow UF_6 + BrF_2^+$$

$$\overline{U + 2BrF_3 \rightarrow UF_6 + Br_2}$$

Plutonium hexafluoride is very much less stable than uranium hexafluoride, and the product of reaction with bromide trifluoride is primarily a lower fluoride which is non-volatile and remains with the fission product residue when the uranium is removed by distillation. There is a very slow reaction of plutonium tetrafluoride with the fluorinating agent to produce sufficient plutonium hexafluoride to present a contamination and purification problem in the separated uranium.

Decontamination factors up to 10^8 for the removal of fission-product activity from uranium can be obtained from a combination of the fluorination and a fractional distillation stage. This compares favourably with the solvent extraction processes, but the complications involved when plutonium has to be recovered adversely affect the economics. A method which has been suggested for plutonium recovery involves the dissolution of the residue in the fluorination vessel in dilute aluminium nitrate solution followed by a fairly conventional solvent extraction. A disadvantage is that the plutonium fluoride is a comparatively small residue spread over the extensive surfaces of the reaction vessel, and large quantities of wash liquor may be required for effective removal. If a more direct method for recovering the plutonium can be found, this fluorination process may find acceptance for the reprocessing of irradiated uranium of short cooling time. Advantages of the method are that if the recovered uranium requires enrichment in the depleted 235 isotope, it can be mixed directly with gaseous diffusion plant material; and if it is to be converted directly to metal for recycling through the reactor the initial reduction to the tetrafluoride is a comparatively easy step.

Another volatilisation process has been suggested for the removal of plutonium from irradiated uranium metal. Plutonium exhibits an appreciable vapour pressure at the melting temperature of uranium, and it has been found possible on a small laboratory scale to obtain up to 99 per cent removal of plutonium by vacuum volatilisation. This is the simplest type of separation system possible, but the materials of construction present a major problem at such high temperatures and the efficient collection of the plutonium requires considerable further investigation. Another disadvantage arises from the fact that although some of the fission products could be predicted to volatilise away from the uranium several of the important neutron poisons would not do so.

PYROMETALLURGICAL PROCESSES

It is to be emphasised that provided facilities can be developed for the refabrication of fuel elements by remote control, separation factors as high as those previously obtained by the solvent extraction processes are no longer necessary. Moreover, if the criterion is to be the reduction of the neutron poison level rather than the reduction of the activity level for handling purposes, the emphasis on the various fission products may be moved and interest in fact now centres primarily around Xe^{135} (neutron cross-section $3 \cdot 2 \times 10^6$ barns) and the lanthanide elements (e.g. Sm^{149}; 66 000 barns).

(i) Oxidative Slagging

A method of removing the fission product neutron poisons from irradiated uranium which has had only a short cooling time is that known as oxidative slagging.

If irradiated uranium fuel is melted in the presence of a very limited amount of oxygen, the most stable oxides will be formed preferentially. These stable oxides, in fact, turn out to be those of the lanthanide fission products and, being less dense, they float on the surface of the uranium. Removal may then be accomplished either by bottom-pouring the metal, so leaving the slag in the melting crucible, or by cooling the crucible contents to room temperature and dissolving the oxides differentially from the surfaces in nitric acid solution. The oxygen necessary for the slagging is supplied partly from the use of ceramic oxide crucibles, by the control of furnace atmosphere, by the addition of oxides to the melt and partly by oxygen impurity levels in the original metal.

Lanthanide removal appears to be readily achieved by this method; what of the other fission products? The free energy of formation of caesium oxide is only 3600 cal/g atom of oxygen compared with 107 000 for cerium, consequently caesium does not enter the oxide slag. It is, however, sufficiently volatile to be removed with the rare gas fission products (the vapour pressure of the metal at 1200°C has been estimated to be about 10^4 mm Hg). Tellurium also has a high vapour pressure at this temperature, but does not volatilise, it appears in the slag due to formation of stable tellurides of the lanthanide elements. If the carbon impurity level in the original uranium metal is of the order of several hundred parts per million, or if the level is raised to this value by deliberate addition of graphite to the melting crucible, much of the zirconium fission product will appear in the slag as the carbide. Ruthenium is not removed from the uranium in this process.

The oxidative slagging process shows promise of being a rapid and compact method of nuclear fuel reprocessing, but much further work is needed before it is likely to be accepted as a practical proposition.

(ii) Fused Salt Extractions

Laboratory work has been carried out on the possibility of separating plutonium from irradiated uranium by contacting the molten metal with molten uranium tetrafluoride. In a system containing an equal weight ratio of salt to metal, 90 per cent of the plutonium is transferred to the salt phase. Although it should be possible to remove all of the plutonium by repeated extractions in this way, the process has not found favour owing to three major disadvantages:

(a) Containment of the salt-metal mixture is difficult in the temperature region required (1200 to 1400°C).

(b) The plutonium is produced in the form of a compound and it would be necessary to add further stages to convert this back to metal.

(c) Significant plutonium losses would occur during the reduction to metal.

(iii) Liquid Metal Extractions

Apart from the possible volatility process for the removal of plutonium from irradiated uranium, potentially the simplest method for reprocessing the latter to enable it to be re-used is a liquid metal extraction. The metallic state is thus preserved throughout and the whole process can be conducted in a relatively small shielded space. The two most promising extractant metals have been found to be silver and magnesium, and the choice between these is difficult since each has rather different advantages and disadvantages.

(a) *Extraction into Silver.* The partition of plutonium has been determined on a tracer scale between molten uranium and several other metals. Silver gives the most favourable distribution coefficient:

$$\frac{\text{mols. Pu/mol. Ag}}{\text{mols. Pu/mol. U}} = 2 \cdot 7$$

Consequently a series of several batch extractions or a continuous column process would remove all the plutonium from the uranium. The disadvantages of silver are its high boiling point, the relatively high solubility of uranium, and a relatively high neutron capture cross-section. The first disadvantage implies that separation of silver from the extracted silver-plutonium mixture would require distillation at very high temperatures under a high vacuum. The solubility of uranium in molten silver is temperature-dependent but is as high as 4 per cent by weight at 1135°C. Thus a single batch extraction with silver can give only about a seventeen-fold increase in the original Pu/U ratio, and this implies multiple extractions.

Finally, the high neutron capture cross-section of silver (60 barns) implies that uranium leaving this process and destined for further irradiation would require the silver content to be kept below about 50 p.p.m. Since the solubility of silver in uranium is 300 p.p.m. at 1135°C, a vacuum distillation on the uranium phase would be necessary to reduce the silver content to an acceptable level.

(b) *Extraction with Magnesium.* Magnesium appears to overcome the disadvantages of silver as an extractant, but has a low distribution coefficient for plutonium. On the same basis as the figure quoted above for silver, the coefficient for magnesium is 0·2. In order to recover 99 per cent of the plutonium it would be necessary to equilibrate the uranium with four successive equal weights of magnesium.

The boiling points and melting points of magnesium, plutonium and uranium have the values:

	m.p.	*b.p.*
Mg	650°C	1106°C
U	1133	∼3900
Pu	660	3240

PLATE I

TRANSPORTING AN OPENED KEG OF CRUDE URANIUM
OXIDE AFTER ARRIVAL AT THE FACTORY

RUNNING A SLURRY OF PRECIPITATED AMMONIUM
DIURANATE INTO A FILTER

CLEANING A URANIUM BILLET OF SLAG AFTER THE
REDUCTION OF URANIUM TETRAFLUORIDE

CHEMICAL PROCESSES AT THE SPRINGFIELDS
URANIUM FACTORY

PLATE II

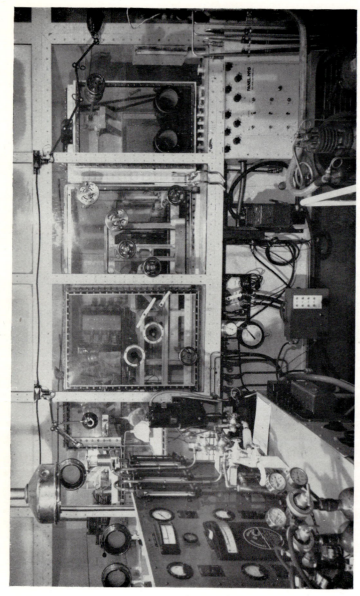

PILOT PLANT FOR THE PRODUCTION OF PLUTONIUM METAL, OPERATED AT HARWELL IN DECEMBER, 1951

PLATE III

THE WINDSCALE PLUTONIUM FACTORY

PLATE IV

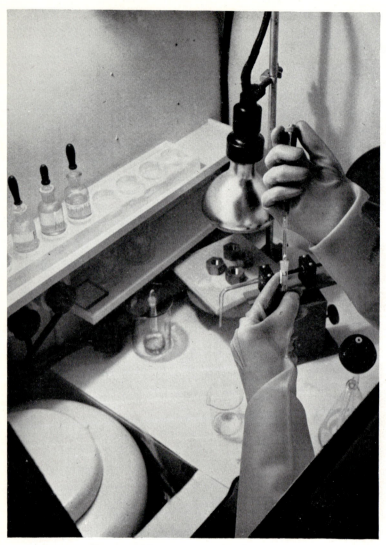

CHEMICAL MANIPULATION IN A GLOVE BOX

PLATE V

GLOVE BOXES IN THE PLUTONIUM LABORATORY, BUILDING 220, HARWELL

PLATE VI

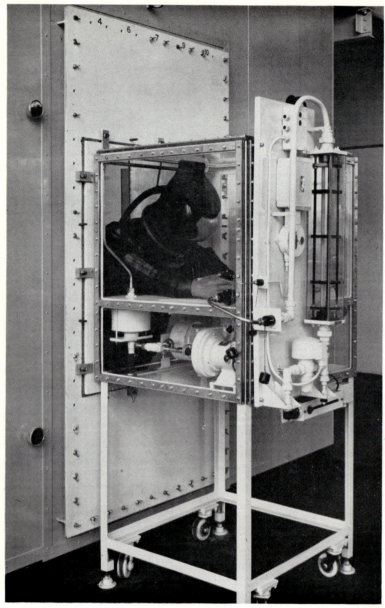

OPERATOR IN A PRESSURISED SUIT CARRYING OUT MAINTENANCE
ON A CONTAMINATED GLOVE BOX

PLATE VII

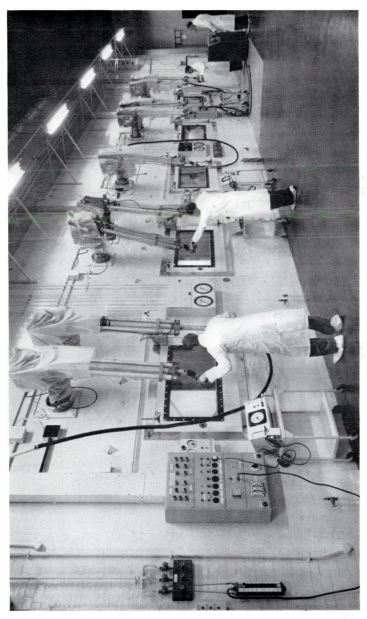

CONCRETE CELLS FOR DISMANTLING HIGHLY GAMMA-ACTIVE EXPERIMENTAL EQUIPMENT, BUILDING 459, HARWELL

PLATE VIII

THE EXTREME DISTORTION OF
URANIUM WHICH OCCURS BOTH ON
EXTENSIVE THERMAL CYCLING

(1) BAR OF URANIUM
BEFORE AND AFTER
THERMAL CYCLING

AND AFTER PROLONGED IRRADIATION

(2) CROSS-SECTION OF
A SAMPLE OF HIGHLY
IRRADIATED URANIUM.
THE FRAME INDICATES
THE ORIGINAL SIZE
OF THE SAMPLE

COULD, EVEN IN THE EARLY STAGES OF
IRRADIATION, CAUSE THE CAN OF
THE FUEL ELEMENT

(3) FUEL ELEMENT FROM THE BEPO REACTOR,
WITH THE CAN CUT AWAY TO SHOW THE URANIUM BAR

TO RUPTURE:

(4) MICROPHOTOGRAPH
OF A RUPTURE IN
A BEPO FUEL ELEMENT

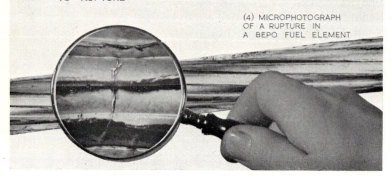

DISTORTION OF URANIUM DURING IRRADIATION

Thus extraction with magnesium at the melting point of uranium must be conducted in a pressurized system, but at 1200°C the excess positive pressure need not exceed 20 p.s.i. This slight disadvantage could be overcome, together with an easing of the containment problem, by operating at a lower temperature made possible by reduction of the uranium melting point to 860°C through alloying with chromium. This leads to further problems, however, since traces of the chromium then appear in the extracted plutonium.

The solubility of magnesium in uranium is about a factor of ten less than the solubility of silver; moreover it has a low neutron capture cross-section. Hence there is no concern over this small amount of magnesium being recycled with the uranium through a reactor. Similarly, the solubility of uranium in magnesium is a factor of twenty lower than the solubility in silver, hence the separation of plutonium from the uranium is a much cleaner process.

This liquid metal extraction process is not very selective for the separation of fission product activity from either the uranium or the plutonium. The volatile fission products, particularly xenon, will be removed into the vapour phase when the uranium is melted. Some of the lanthanide fission product neutron poisons will transfer to the molten magnesium along with the plutonium, but the remainder would be expected to form a solid scum by formation of oxides and carbides from impurities in the system; this scum would possibly introduce complications into the process of separation, perhaps a filtration step might become necessary. About 20 per cent of the ruthenium-niobium-zirconium group of fission products would enter the magnesium phase, the remainder staying with the uranium.

One of the factors which remain to be solved is the provision of a suitable containing material for the liquid metal extraction system. Tantalum is the most promising metal for this purpose, but since its solubility in uranium is 0·3 per cent at 1150°C it would not be capable of long-term use and failure would eventually occur. Ceramic coatings may be of considerable value to cut down this corrosion rate, but might in turn aggravate the formation of solid fission-product slags.

SEPARATION OF URANIUM FROM IRRADIATED THORIUM

The separation processes which have been discussed are concerned with the removal of plutonium and the fission products from irradiated uranium fuel rods. With the advent of two-zone breeder reactor designs it has become necessary to consider the separation of

uranium, thorium and the fission products. Although it is possible
to use either U^{238} or Th^{232} in a breeder blanket, the superior
nuclear properties of U^{233} make it highly desirable to prefer the
thorium cycle.

We have mentioned the problem of cooling time for irradiated
uranium separation processes. The same problem occurs in an
even more complex form in irradiated thorium. This may be
illustrated by reference to Fig. 11, which shows the principal nuclear
reactions which occur when thorium is introduced into a reactor.
Each of the end-products shown in the Figure also undergoes
further reactions: thus the U^{234} can capture a further neutron to
give the fissile isotope U^{235}, and the U^{232} decays by alpha emission
to Th^{228} which is an alpha emitter with 1·9 y half-life.

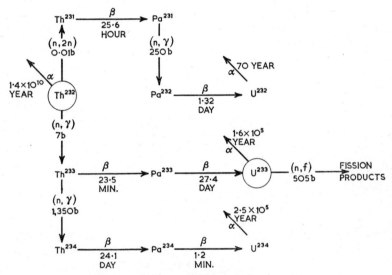

FIG. 11. PRINCIPAL NUCLEAR REACTIONS OCCURRING DURING THE
IRRADIATION OF Th^{232}

The primary aim must be to extract the maximum possible
amount of U^{233}, but this is complicated by the rather long half-life of
Pa^{233}. If the separation process is specific in extracting uranium,
and leaves protactinium with the thorium, relatively long cooling
times are required to obtain the full yield. Otherwise it becomes
necessary to complicate the extraction method to include separation
of Pa^{233} and its storage for ultimate decay to the desired uranium
isotope. Short cooling times can lead to additional problems,

however, if the separated thorium and uranium are to be handled by conventional means without radiation shielding. Th^{234} will remain with the purified thorium after short cooling times and will give rise to considerable levels of radio-activity, particularly since its Pa^{234} daughter emits both beta and gamma radiation of considerable energy. A further difficulty arises from the U^{232} which would be separated with the uranium after short cooling times. This isotope decays by alpha emission to give Th^{228} and further radioactive daughters; thus activity would build up in the uranium for some time after separation from the thorium and it might then become necessary to reprocess the uranium to lower the activity level. On the other hand intermediate cooling times would imply the separation of larger amounts of Th^{228} in the thorium with subsequent build-up of radioactivity.

These considerations are important only as long as the handling of the separated thorium and uranium during the processes for their reconversion to metal rods must be done without radiation shielding. If remote techniques are developed for all of the intricate processes involved, then full advantage could be taken of the beneficial economic effects of short cooling times. There is a wide scope both here and in the plutonium-uranium separation processes for the fullest use of automation provided that the latter can be developed in a relatively cheap and reliable form.

The only uranium-thorium separation process upon which considerable development has been undertaken is that of solvent extraction. The primary step in the process is then the dissolution of thorium in an aqueous medium. Unlike uranium, the solution rate of thorium in nitric acid is very slow owing to the formation of a protective film of thoria. It has been found that the addition of a small quantity of fluoride ion (e.g. as ammonium fluoride) to the nitric acid allows penetration of the oxide film to take place, but the amount added has then to be a compromise between increasing the solution rate and increasing the corrosion rate of the stainless steel containing equipment.

Tributyl phosphate diluted with kerosene has been found to be suitable for the solvent extraction stages in counter-current columns. Uranium is preferentially extracted from the nitric acid solution in the first column and is then washed back in a second column, an increase of 10^5 in the uranium/thorium ratio being obtainable over these two stages. Separation of the thorium and fission products is achieved in a third column by extraction into a much more concentrated solution of TBP in kerosene. At this stage, the fluoride ion added in the dissolving step becomes a nuisance and it is necessary to add aluminium nitrate in order to form a stable fluoride complex and so prevent the precipitation of thorium tetrafluoride at the dilute end of the extractor.

One final remark which should be made in closing is that the economy of operation of any separation plant of the types considered here is directly affected by its throughput. The higher the throughput, the lower the cost of the product. Hence it is unlikely that large numbers of processing plants will be built; rather there will be a tendency to make very few plants serve the diverse requirements of many reactors. Consequently the designer must be sure he has picked a really good process and that it is sufficiently versatile. A very considerable amount of detailed development work is needed to be able to decide the relative merits of the possible processes.

FURTHER READING

Benedict and Pigford. *Nuclear Chemical Engineering.* McGraw-Hill Book Co. Inc., 1957.

Bruce, Fletcher, Hyman and Katz. *Progress in Nuclear Energy,* Series III. Process Chemistry, Vol. 1. Pergamon Press, 1956.

United Nations. Proceedings of the First International Conference on the Peaceful Uses of Atomic Energy, Geneva, 1955. Vol. 9, *Reactor Technology and Chemical Processing.* Second International Conference on the Peaceful Uses of Atomic Energy, Geneva, 1958. Papers 307, 1795, 3388.

CHAPTER 6

THE HANDLING OF RADIOACTIVE
MATERIALS

PRIOR to 1940 only comparatively small amounts of radioactive materials were available. These were obtained either from natural sources, for example radium, or by irradiation of naturally-occurring isotopes in the beam of a particle accelerator. This method produced only trace amounts of activity. With the operation of nuclear reactors radioisotopes became available on an ever-increasing scale. Not only are the by-products of the operation of a reactor highly radio-active but specific radioisotopes can be made of high activity by the irradiation of suitable naturally occurring elements. The rapid rise in the availability of radioactive materials may be seen by comparing the total amount of radium in use in the United Kingdom—of the order of hundreds of curies—with the annual output of a single low-powered reactor. A 10 MW natural-uranium research reactor produces in one year about 10^6 curies of long-lived beta and gamma active fission products and 200 c of alpha-active Pu^{239}. It can also produce, by the irradiation of suitable materials, thousands of curies of various radioactive isotopes.

Fortunately, prior to the operation of nuclear reactors and the subsequent production of large amounts of radioactivity the potentially hazardous nature of radioactive materials was fully appreciated. Wide, and occasionally fatal, experience has been gained in the operation of therapeutic X-ray equipment and in the use of radium both in medicine and as a basis for luminous paints. From the very beginnings of the atomic energy industry adequate precautions have therefore been taken in the handling of radioactive materials. These have been so successful that in the United Kingdom there has been no case of injury arising from exposure to radioactive materials. This safety consciousness also reflects back upon the general accident rate, which in the U.K.A.E.A. is substantially lower than for the chemical industry as a whole.

The need for stringent health precautions in the handling of radioactive materials has given rise to a new branch of industrial hygiene known as health physics. The responsibilities of this branch include the determination of maximum permissible radiation and ingestion doses, the measurement of exposures received by operating staff and the assessing of the potential radiological hazards of proposed experimental procedures and plant operations, besides the day-to-day checking of laboratories and plant in which radioactive materials are handled.

NATURE OF THE HAZARDS INVOLVED

The consensus of opinion is that biological damage in the human body arises primarily from radiation-induced chemical reactions within the living cell. Any form of ionising radiation is therefore potentially dangerous. Two important facts must be borne in mind when considering the biological effects of radiation. Firstly, a level of exposure can be set such that no appreciable bodily harm would be expected to an individual during his lifetime. For occupationally exposed persons, the average maximum permissible level for exposure of the whole body to X- or gamma-radiation is 0·1 röntgen per week, subject to a total of 50 röntgens by the age of 30 years (for genetic reasons) giving a total of 200 röntgens by the age of 60 years. Such a total dose received over a short period of time would cause severe illness and, in some individuals, could be fatal. Secondly, multiplying cells are particularly sensitive to radiation. For this reason the most sensitive parts of the body include the bone marrow, in which the blood-cells are produced, and the reproductive organs. For this same reason cancerous growths, where uncontrolled multiplication of cells is occurring, and which may be more susceptible to radiation damage than the surrounding normal tissue, can sometimes be successfully treated by X-ray and gamma-ray therapy.

It is important to distinguish between two types of radiological hazard. Firstly, sources of penetrating radiation external to the body cause ionisation within the body. Only radiations which penetrate the skin are important here. These include X-rays (from X-ray machines), neutrons (from, e.g. a nuclear reactor), high energy particles (from accelerators) and beta or gamma radiation (from radioactive materials). The particles emitted by alpha-active materials are not very penetrating and so do not constitute an external radiation hazard. Secondly, intake into the body of any radioactive isotope can constitute a serious hazard, and the problem is particularly acute when handling isotopes emitting alpha or beta particles.

Two widely different types of problem then face the chemist who is handling radioactive material, depending on whether the material emits the penetrating gamma radiation or the less penetrating alpha or beta particles. In the following sections these two problems will be considered individually, although instances do arise in which both problems are met simultaneously, as, for example, in the handling of irradiated plutonium.

GAMMA-ACTIVE MATERIALS

The intensity of the radiation, and hence the dose received, from an unshielded gamma-source depends upon the strength of the source in curies (that is the total rate of radioactive disintegration, 1 c being $3 \cdot 7 \ 10^{10}$ disintegrations/sec), and upon the energy of the gamma-rays. The intensity decreases by the inverse square law with increasing distance from the source. The simplest means of obtaining protection from a gamma source is therefore to work at a distance. This, however, is only practicable for the smallest sources, as indicated in Table 15.

TABLE 15.—THE MAXIMUM WORKING TIMES PER WEEK
AT VARIOUS DISTANCES FROM FISSION-PRODUCT SOURCES OF
$1 \cdot 0$ AND $0 \cdot 01$ CURIES

Working distance (ft)		$\frac{1}{2}$	1	2	5	10	50
Maximum working time per week (approx.)	$\begin{cases} 1 \cdot 0 \ c \\ 0 \cdot 01 \ c \end{cases}$	15 sec 25 min	1 min $1\frac{1}{2}$ h	4 min 7 h	25 min 40 h	$1\frac{1}{2}$ h 150 h	40 h 168 h

Thus for safe full-time working a minimum distance of 50 ft is required for a 1 c source. This distance is impractical for work which involves any degree of careful manipulation, so that some other form of protection is necessary. This can be provided by interposing a barrier of dense material between the source and the observer. The degree of attenuation of the gamma radiation by a shield of given thickness is greater the higher the density of the shield, but it depends also upon the energy of the emitted gamma rays. The most commonly used shielding materials are lead, steel, concrete and, for certain limited applications, water. In Table 16 the approximate thickness of material which is needed to attenuate the gamma intensity by a factor of ten is given for a selection of sources and shielding materials.

The precise calculation of the thickness of shielding required for a given source is not an easy matter, especially when the source is of large volume as, for example, is the core of a nuclear reactor. Such

TABLE 16.—THICKNESS OF SHIELD, IN INCHES, WHICH WILL
ATTENUATE THE GAMMA INTENSITY BY A FACTOR OF TEN
(ASSUMING A BROAD BEAM OF GAMMA RAYS)

Source	Maximum γ-energy (MeV)	Water (1·0 g/cm³)	Concrete (2·3 g/cm³)	Steel (7·9 g/cm³)	Lead (11·4 g/cm³)
Na²⁴	2·7	34	14	4·8	2·5
Co⁶⁰	1·3	26	11	3·6	1·8
Cs¹³⁷	0·7	24	9	2·9	1·0
I¹³¹	0·4	23	7·5	2·1	0·6

calculations are very necessary, however, for a considerable increase
in expenditure can result from overshielding.

Having provided protection for the operators, some means of
manipulation and viewing are required. The methods of achieving
these depend upon the nature and thickness of the shield. For
sources of up to 1000 c, walls of lead of up to 10 in. in thickness,
may be used in conjunction with a series of small suitably-placed
windows of a dense lead-glass and tongs operating through ball-
and-socket joints in the wall. For sources of greater than 1000 c,
and in some cases for smaller sources, it often proves cheaper to
shield with concrete of up to several feet in thickness. Because of
the thickness of the wall master-slave manipulators are used, and
the windows must necessarily be large in area to give an adequate
field of view. Two types of window are used for concrete-shielded
installations. One consists of multiple sheets of plate glass, which
are deliberately separated by about 0·01 in. in order to prevent the
formation of Newton's rings, and are immersed in paraffin to reduce
reflection at the surface of each plate. The other type consists of a
solution of zinc bromide in water. This salt is used because it is
reasonably stable to radiation and is very soluble in water, giving a
density of 2·5 g/cm³. The shielding properties of such a window
are therefore about the same as the concrete of the wall in which it is
located.

The use of each type of window—glass or solution—presents its
own chemical problems. Under the intense radiation glass darkens
considerably, but this can be prevented by the incorporation of a
small proportion of ceric oxide in the glass. A solution of zinc
bromide liberates free bromine on exposure to radiation. The
presence of small amounts of a reducing agent, such as hydrazine
hydrobromide or hydroxylamine hydrochloride, has a stabilising
effect by reducing any free bromine back to bromide. At very high

exposure rates, exceeding $5 \cdot 10^4$ r/h, radiolytic decomposition of the water is so rapid that bubbling of the solution impairs the vision. For handling sources of the highest activity, therefore, it is necessary to protect the window from the intense radiation by means of several layers of ceria-stabilised plate glass. In constructing a zinc bromide window, salt of the highest purity must be used and corrosion of the structure of the window must be avoided, for even a few parts per million of, for example, iron can seriously discolour the window. For this reason a polyester is preferred to metal for the construction of the window-tank.

ALPHA- AND BETA-ACTIVE MATERIALS

Both alpha and beta particles travel only short distances in solid materials before being brought to rest, the ranges being of the order of 50μ and 1 cm respectively for high energy particles. A rubber glove for alpha emitters or a sheet of perspex for beta emitters therefore give adequate radiation protection for the operator. However, practically all radioactive isotopes are absorbed by the human body and because of the short range of the emitted alpha or beta particles all of their energy is deposited within the organ involved. Alpha and beta emitters therefore constitute a serious internal hazard. They may enter the body orally, through the respiratory organs, through a skin lesion, or even by direct absorption through the skin. Following intake into the body

TABLE 17.—BIOLOGICAL HAZARD OF ALPHA AND BETA
EMITTERS FOR CONTINUOUS OCCUPATIONAL EXPOSURE

Isotope	Radiation	Critical Organ	Maximum Body Burden		Maximum Concentration	
			(a) (μc)	(b) (μg)	In Drinking Water ($\mu c/l$)	In Air ($\mu c/l$)
C^{14}	β	Fat	260	800	3	10^{-2}
Na^{24}	β, γ	General	15	8.10^{-6}	8	2.10^{-3}
Sr^{90}	β	Bone	1	10^{-2}	8.10^{-4}	2.10^{-7}
I^{131}	β, γ	Thyroid	0·6	1	6.10^{-2}	6.10^{-6}
Cs^{137}	β, γ	Muscle	98	15	2	2.10^{-4}
Mixed Fission Products	β, γ	—	--	—	10^{-4}	10^{-6}
Po^{210}	a, γ	Spleen	0·04	8.10^{-6}	3.10^{-2}	5.10^{-7}
U^{233}	a, γ	Bone	0·04	4	0·15	3.10^{-8}
Pu^{239}	a, γ	Bone	0·04	0·5	6.10^{-3}	2.10^{-9}
Am^{241}	a, γ	Bone	0·06	2.10^{-2}	0·2	4.10^{-8}

attachment to tissue may ensue. For example, Pu^{239} is deposited in the bone, particularly near its outer and inner surfaces. In the latter position it is well situated to irradiate the bone marrow, and this can result in disorders of the blood such as leukaemia, or in the formation of bone tumours.

On ingestion of a quantity of radioactive material, a certain proportion is absorbed into the body, a large fraction of the remainder usually being rapidly excreted. This is followed by the slow excretion of the radioactive isotope. To this process a biological half-life can be assigned for any particular element or any particular organ. Knowing this half-life, the radioactive half-life, the energy of the radiation, the site of deposition and the degree of retention when breathed or swallowed, the dose received by the organ involved can be estimated. It is then possible to calculate the maximum permissible values for the total body burden; the permissible concentrations in air or drinking-water can then be established. Some typical values are summarised in Table 17.

The quantities involved are minute and it is therefore imperative that these materials be handled in such a way that the inevitable spread of contamination is adequately controlled. Fortunately the same radiation which constitutes the hazard facilitates the location and estimation of any untoward contamination by means of sensitive radiation monitors.

The handling of alpha- or beta-active materials is therefore carried out in sealed boxes normally constructed of metal with Perspex working faces, and provided with arm-length rubber gauntlets sealed into the Perspex at the shoulder for manipulation. The boxes are maintained at a pressure slightly less than atmospheric so that any leaks are inwards. Small amounts of activity (of the order of millicuries) may be handled in fume hoods having a good air-flow. Even smaller quantities (of the order of microcuries) in solution, where the possibility of spread of activity is minimised, may be handled in a well ventilated laboratory. In both cases the hands of the operator are protected from contamination by surgical rubber gloves.

An additional problem arises in the handling of large sources of certain beta-emitters, for when high energy beta-particles are stopped in matter X-rays, known as *brehmsstrahlung*, which are more penetrating than the original beta radiation, are produced. When the intensity of the brehmsstrahlung is sufficient to constitute a radiation hazard adequate shielding must be provided. Manipulation in this case may require the use of tongs working in ball-and-socket joints.

Certain compounds of highly alpha-active isotopes also present a radiation hazard from neutrons. Certain light nuclei, for example beryllium or fluorine, capture high energy alpha particles and simultaneously emit a fast neutron:

$$F^{19} + a^4 \rightarrow Na^{22} + n^1$$

Thus, a sample of the fluoride of an alpha-active isotope containing more than a few curies of the isotope emits a significant flux of fast neutrons, which must be attenuated by means of blocks of paraffin wax containing boron. The hydrogen in the wax brings about the slowing-down of the neutrons by collision until the neutron energy is low enough for strong capture in the boron to occur.

SOME TYPICAL LABORATORY INSTALLATIONS

To illustrate some of the problems which arise when carrying out chemical operations on active materials three typical installations will be described. These are boxes for handling plutonium, a sealed lead-shielded cell for the handling of mixtures of alpha, beta and gamma emitters and a concrete general-purpose cell for handling gamma emitters.

BOXES FOR HANDLING PLUTONIUM

In connection with the homogenous aqueous reactor, which is discussed in Chapter 7, it was necessary to investigate the physical properties of solutions of a range of plutonium salts at temperatures up to 300°C. The major item of interest was the thermal stability of the solutions, and this was determined by slowly heating a small sample, sealed in a glass ampoule, in an electric furnace and observing any phase changes which occurred. This involved the preparation and analysis of salts of tri-, tetra- and hexavalent plutonium, the making-up of solutions of known strength and the analysis of solutions by gravimetric, volumetric and counting techniques.

The various procedures were carried out in a suite of nine boxes of the standard Harwell pattern. This type of box is of mild steel, 3 ft wide, 2 ft deep and 2 ft 6 in. high, and the top and working face are of Perspex. The box is sealed, and maintained at a pressure slightly below atmospheric by means of an air ejector operating through a filter unit. Materials are transferred into or out of the box without breaking the seal by means of P.V.C. bags fitted over a 6 in.-diameter flanged hole in the side of the box. The end of the bag, containing the material being transferred, is sealed by means of a radio-frequency heat sealer, and the sealed portion of the bag is cut off along the seam.   The main features of a standard glove box are shown in Fig. 12.

Evaporation of Solutions.   Although the boxes are maintained at a slight negative pressure with respect to the laboratory, the air-flow into the box is deliberately kept as low as practicable in order to minimise the escape of activity in the event of the failure of the extract.   Evaporation of solutions, and particularly of acid solutions, would quickly result in the corrosion of equipment within the box.   A glass funnel connected, via a filter, to an air ejector and inverted over the hotplate on which the evaporation was being carried out effectively prevented fumes from diffusing throughout the box.

Ozonisation.   Oxidation of plutonium solutions was achieved by means of ozonised oxygen, prepared by a silent discharge apparatus outside the glove box.   Release of the excess ozone directly into the box would result in the rapid perishing of the rubber gloves.   The effluent gases from the oxidation were therefore led through a P.V.C. tube to a side-arm on the evaporation hood described above.

Vacuum Filtration.   Suction for filtration was provided by an air ejector outside the box.   The partial vacuum was transmitted to a Buchner flask by means of pressure tubing through a stub of pipe welded through the wall of the box.   When the filter pad became dry there was a danger of evacuating the box to a point where the panels would collapse inwards.   This danger was obviated by using an air ejector of low flow-capacity and providing a small leak of air into the box whenever filtration was carried out.

Vacuum System.   For ease of maintenance, and in order to save valuable glove-box space, the rotary and diffusion pumps of the vacuum system were mounted outside the glove box.   The exhaust of the rotary pump was returned to the glove box to avoid evacuation of the box in the event of a leakage in the vacuum system and to ensure safe disposal of the exhaust should this be contaminated.

Glass and silica ampoules were sealed under vacuum in this box. The use of an oxy-coal gas torch in a closed box presents a considerable fire hazard and involves the risk of an explosion.   A carbon arc, operated from a 200 V d.c. supply, was therefore used to provide the intense local heating for sealing-off the ampoules.

Ignition.   Again the use of an oxy-coal gas torch for ignition is highly undesirable and large-scale ignitions were therefore carried out at temperatures of up to 1200°C in an electric furnace employing silicon carbide heating elements.   To cool the box, air was drawn through it at a rate of 80 ft³/min and exhausted to the build-

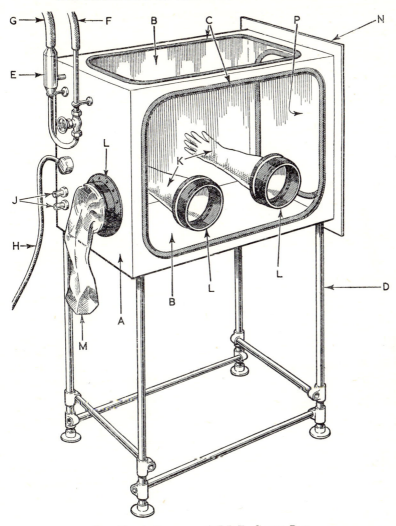

FIG. 12. A STANDARD A.E.R.E. GLOVE BOX

A. Mild-steel shell
B. Perspex panels
C. Rubber sealing strip
D. Rubular steel stand
E. Compressed-air ejector
F. Compressed-air supply
G. Exhaust to the building extract system
H. Power supply cables

J. Inlet/outlet for cooling-water circuits, etc.
K. Rubber or neoprene gauntlets
L. Flanged glove-ports
M. P.V.C. bag for transfer of materials into and out of box
N. Flange for attaching box to active maintenance area
P. Removable steel panel

ing extract by means of a fan. As an additional precaution the box windows were of a special heat-resisting variety of Perspex.

Handling. Very little restriction is placed on the operator by working in gloves. Provided that the layout of the box is well designed, and the equipment is within easy reach, any operation can be carried out. The wearing of gloves only slightly impairs the sense of touch (see Plate IV). It is quite feasible, for example, to seal off fine X-ray capillaries by hand in a glove box. The major delays in the performing of experiments in glove boxes arise in the initial design stages, and, during operation, in the transfer of materials into and out of the box.

THE PLUTONIUM LABORATORY (see Plate V)

The glove boxes described above are located at Harwell in a building which was specially designed for the handling of alpha-active materials. Two basic types of laboratory facilities are available. Firstly, there are suites consisting of laboratory, office and vestibule; the laboratory contains fume-hoods suitable for the handling of small amounts of activity. Secondly a wing of the building is devoted entirely to the accommodation of free-standing glove boxes. These are located in a large hall and receive the necessary services of compressed air, water, electricity and extract from a service floor above. In order to make any major modifications to the equipment within a glove box, the box is disconnected from its services and wheeled to the working face of an active maintenance area located at one end of the hall. This area, which is provided with a complete workshop, is considered to be contaminated, and may only be entered by personnel wearing a pressurised suit of rubber or P.V.C. A flange round one end of the box is sealed against a corresponding flange on the wall of the maintenance area. Operators within the area gain access to the interior of the box by removing both a door in the wall and the end panel of the box (see Plate VI). After the completion of the operation the panel is replaced and decontaminated, and the box is returned to its place in the hall for further use.

Several special services and facilities in the building are provided with a view to minimising the spread of contamination both during normal operation and in the event of an emergency. Protective clothing, in the form of laboratory coat and either special shoes or canvas overshoes, is worn in the laboratory, and ample facilities for monitoring both hands and clothing are provided. The air in the building is circulated by an extensive plant, the general direction of air-flow being from the least active to the most active areas. The level of contamination of all exposed surfaces is checked regularly by health physics officers, and any contamination is removed. A continuous check is also made of the level of air-borne activity in the working areas and in the air discharged from the building.

In the event of an accident, and especially of fire, speed is essential, and push-buttons liberally distributed call the health physics officers and the fire station directly, besides giving warning to the occupants of the area concerned in the incident.

A LEAD-SHIELDED CELL

As an example of an alpha-beta-gamma facility, a cell designed for chemical analysis of aqueous uranium solutions which have been irradiated in one of the reactors BEPO or DIDO will be described.

The solution, which has been irradiated in a small metal autoclave, is brought from the reactor in a heavy lead flask and transferred into the cell by remote handling techniques. Because solutions of Pu^{239} or U^{233} are used in some experiments it is essential that the cell should not only provide adequate shielding for the gamma-radiation of the fission products but also be hermetically sealed. The activity level anticipated was up to 20 c for intermittent operation and 1 c for continuous operation. Lead was selected as the shield material, and a thickness of 4 in. was considered to provide sufficient attenuation.

The cell is depicted in Fig. 13. It consists of a hermetically-sealed shell of mild steel with suitably placed Perspex windows, standing on a concrete plinth and surrounded by a wall of interlocking 4 in. lead bricks. As is usual with lead walls, manipulation is carried out by means of tongs working in ball-and-socket joints in the wall. The tong-heads are hermetically sealed to the inner steel shell by means of flexible gaiters of P.V.C. A removable lead door at one end gives access to a P.V.C. bag through which items may be posted into or out of the box by the technique already described.

The operations which are carried out in the box are as follows:

(i) Removal of solution from the autoclave.
(ii) Opening of the autoclave.
(iii) Preliminary examination of the interior of the autoclave and the corrosion specimens.
(iv) Weighing of the specimens.
(v) Determination of the pH of the solution.
(vi) Chemical analysis of the solution.

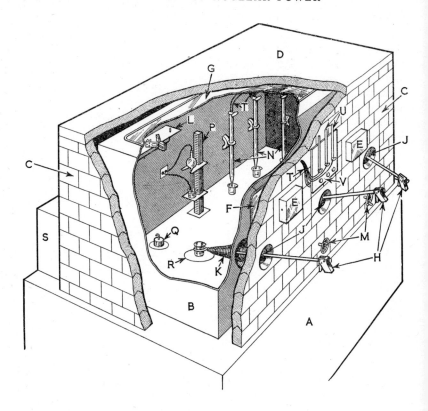

FIG. 13. A LEAD-SHIELDED CELL

A. Concrete plinth
B. Mild steel inner shell
C. Walls of interlocking 4″ lead bricks
D. Roof of lead bricks laid on steel sheet
E. Lead-glass window
F. Perspex windows for viewing
G. Perspex window to admit light from lamps in the void under the lead roof
H. Remote-handling tongs
J. Ball joints
K. P.V.C. gaiters, sealed to tong-head and box
L. Tong-head changing station
M. Hand-wheels operating mechanisms within the box

N. Ion-exchange and solvent extraction columns, attached to box by magnets
P. Rack and pinion carrying pipette, pH electrode, etc.
Q. Cover of storage-chamber
R. Insert of stainless steel, with magnetic stirrer beneath
S. Lead box, containing spectrophotometer call
T. Fine polythene tubes for compressed air, suction and transfer of reagents
U. Stock solutions of reagents
V. Suck-blow control for transfer of liquids

The solution is removed by inverting the autoclave over a beaker and heating the body of the autoclave in a small electric heater; this forces solution down the capillary tube which is built into the head of the autoclave. Rinsing is carried out in a similar manner. It is then opened either by cutting in a mechanical pipe-cutter or by unscrewing the head-nut by means of a box spanner with a long extension which passes out through the top of the box.

The balance on which the sample is weighed is located in an unshielded glove box on top of the cell, mainly to protect the balance from the moist atmosphere of the box and to make possible routine maintenance. One pan of the balance is located in a small Perspex box within the cell and is suspended on a fine thread, which passes down a $\frac{1}{2}$-in. diameter pipe from the balance box. An intrascope, inserted through a small hermetically-sealed hole in the side of the cell, is used to view the interior of the opened autoclave and corrosion samples.

The solution, being of higher activity than can be safely handled continuously in the box, is stored in a small coffin, with extra lead shielding, located in the floor of the cell. Chemical analysis of aliquots of this solution involves preliminary separation of some of the elements present and this is achieved by ion-exchange and solvent-extraction techniques, followed by determination of the cations present by means of a Spekker photoelectric colorimeter. An apparatus for coulometric analysis is also included in the cell. In designing chemical equipment for use within such a cell two basic limitations must be borne in mind. Firstly, the restricted number of degrees of freedom of the tongs imposes a limit on the types of movement which can be made. For example, rotation is only possible about the axis of the tong, so that such items as glass stop-cocks must be directed along this axis. Secondly, any piece of equipment which is liable to failure during operation must be easily replaceable. The ion-exchange and solvent-extraction columns are therefore held in place on the steel walls of the box by permanent magnets, rather than by rigid screws and brackets.

Solutions are transferred wherever possible through fine polythene tubes either by suction or by blowing. Stirring is accomplished either by bubbling a stream of air through the solution or by a magnetic stirrer. An electrically-rotated permanent magnet is located beneath a disc of non-magnetic stainless steel in the floor of the inner box, and the solution to be stirred, containing a polythene-covered bar magnet, is placed above it. Solutions are pipetted in aliquots of 0·1 to 0·5 ml. in automatic pipettes, which, when they are filled by suction and allowed to drain under gravity to the calibration constriction, deliver the aliquot on blowing out the solution.

The optical cell of the colorimeter is mounted behind lead shielding outside the handling box. The light source of the instrument is located outside the lead shield and the light beam passes through a small hole in the shield on to a modified optical cell and is reflected back out of the shield on to the photoelectric detector. Solution is transferred from the main part of the box along a polythene tube leading directly into the optical cell, and after the measurement is returned to the box by suction. The optical cell is cleaned and dried in a similar manner.

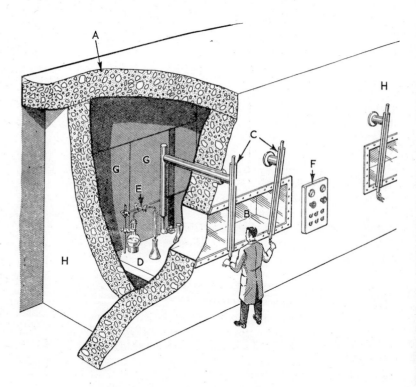

FIG. 14. A CONCRETE-SHIELDED CELL

A. Biological shield (concrete)
B. Window (tank of $ZnBr_2$ solution)
C. Pair of master-slave manipulators
D. Work-bench
E. Chemical apparatus

F. Control panel for equipment within the cell
G. Sliding doors (steel) giving access from rear of cell
H. Neighbouring cells

A CONCRETE CELL

When concrete is used as the shielding material the thickness of the wall precludes the use of ball-and-socket tongs, and recourse has to be made to the more expensive but much more versatile master-slave manipulators (see Plate VII). A typical concrete cell is shown in Fig. 14. The movements here are much less restricted than with the swivel tong, the tong-head being at least as mobile as the human wrist. By extension and flexion of the supporting arm the head can reach almost any point within the cell. A further feature of this type of facility is the large window size. A large window is necessary because of the greater mobility of the operator when using master-slave manipulators. Unfortunately the increased working distance (of up to 10 ft) renders the estimation of depth by binocular vision somewhat uncertain.

A master-slave manipulator can perform almost any operation requiring only two fingers of each hand, and therefore special techniques are not so vital here as they are with the swivel-tongs used in lead cells. Many of the techniques used in the lead cells, such as the transferring of liquids by suction rather than pouring can, however, be usefully employed in a concrete cell.

Although the chemical manipulation of radioactive materials presents many formidable problems, these have largely been overcome by ingenious engineering design of remote handling equipment. Considerable ingenuity and foresight on the part of both chemist and engineer are required, not only in designing suitable equipment, but also in modifying even the simplest of techniques, such as the pouring of liquids, to meet the new conditions. In fact the whole field of remote handling is a veritable gadgeteer's paradise.

FURTHER READING

Barnes and Taylor. *Radiation Hazards and Protection*, George Newnes Ltd. 1958.
Medical Research Council Memorandum. *Introductory Manual on the Control of Health Hazards from Radioactive Materials.* Her Majesty's Stationery Office (1958).
U.K. Atomic Energy Authority. *Glove Boxes and Shielded Cells.* Butterworths Scientific Publications (in the press).
U.S. Atomic Energy Commission. *Chemical Processing and Equipment.* McGraw-Hill Book Co. Inc., 1955.
U.S. Atomic Energy Commission. *Hot Laboratory Equipment.* Technical Information Service, Washington, April 1950.

CHAPTER 7

WATER-COOLED NUCLEAR REACTORS

THERE are several different ways in which nuclear reactors may be classified. They may be known as thermal, intermediate, or fast, according to the energy of the neutrons which are used to maintain the chain reaction. There are homogeneous and heterogeneous reactors—the difference lying in the distribution of the fuel, moderator and coolant. This chapter will describe the chemical problems associated with reactors employing light or heavy water as coolant. It is convenient to discuss water-cooled reactors collectively, since many of the problems are common to them all. This group may be subdivided as follows:

>   *Low Temperature.* Used for research or for plutonium production. Solid fuel elements separated from the water by a canning material.
>   *Pressurised Water.* For power production. Heat is extracted from the reactor core by pumped water under high temperature and pressure.
>   *Boiling Water.* As above, but heat removal is accomplished by allowing the water in the core to flash off as steam. Less pumping power is required, but the physics of boiling presents engineering and nuclear problems.
>   *Homogeneous Aqueous.* The fuel, moderator and coolant are intimately mixed and are circulated together in the heat removal circuit. Some designs have considered a boiling system for heat removal.

All of these types may be designed for use with either light or heavy water, the resultant systems then having quite different characteristics.

The first water-cooled reactors were built in the United States; CP–3, a heavy-water moderated and cooled reactor, was completed in 1944 at the Argonne National Laboratory and had a power level of 300 kW. The much larger plutonium production reactors constructed at Hanford in 1945 are cooled by water from the Columbia river. Since that time American reactor designs have tended to continue along the lines of water cooling, whereas the United Kingdom designs have tended to favour gas cooling following the construction of the early gas-cooled reactors such as BEPO at Harwell and the plutonium production reactors at Windscale. Nevertheless some designs of water-cooled power reactors are under investigation in the United Kingdom (for instance for submarine propulsion) and in several other countries. The first Russian commercial atomic power station was of this type.

There are some problems of chemistry which are common to all water-cooled reactors. Since, however, the homogeneous reactors have many additional problems specific to the use of a liquid fuel, the simpler heterogeneous reactors will be discussed first.

HETEROGENEOUS WATER-COOLED REACTORS

The use of water in a nuclear reactor gives rise to two main chemical problems: radiation decomposition of the water and corrosion of metallic constructional materials at elevated temperatures. These problems are essentially the same whether light water or heavy water is utilised.

It has been known for many years that water may be partially decomposed by the passage of radiations through it. For instance, Cameron and Ramsay in 1907–8 studied the decomposition by alpha particles from radon. The decomposition of water by other types of radiation has been the subject of intensive study by many groups of workers and it is only recently that a clear picture of the mechanism has emerged. Even as late as 1947 when the NRX heavy-water moderated and cooled reactor was brought into operation at Chalk River there had been speculation about whether the coolant would be stable under the intense radiation. Not until the reactor was operated was it confirmed that the net decomposition rate of the heavy water could be made extremely low, even at high power operation, provided that suitable water conditions were maintained. The reason for this will appear if we look a little more closely at the present state of knowledge of the radiation chemistry of water.

It is believed that irradiated water may decompose in two different ways:

$$2H_2O \rightarrow H_2 + H_2O_2 \quad (F)$$
$$H_2O \rightarrow H + OH \quad \quad (R)$$

With radiations of low ionisation density, such as gamma rays, the *radical reaction* (R) predominates, whereas for highly ionising radiations such as alpha and beta particles and fission fragments the molecular reaction (usually known as the *forward reaction*, F) becomes more important. In a nuclear reactor, with its mixture of radiation types (gamma rays and energetic protons produced by neutron interactions) both of these reactions appear to proceed simultaneously. The radicals H and OH are extremely reactive species and combine rapidly with the products of reaction (F):

$$H_2 + OH \rightarrow H_2O + \dot{H}$$
$$H_2O_2 + H \rightarrow H_2O + \dot{O}H$$

If we add these two reactions together, we obtain

$$H_2O_2 + H_2 \rightarrow 2H_2O$$

—the reverse of the original decomposition reaction (F). The very low net rate of heavy water decomposition observed in the NRX reactor, and later in DIDO, was therefore the result of the yield of

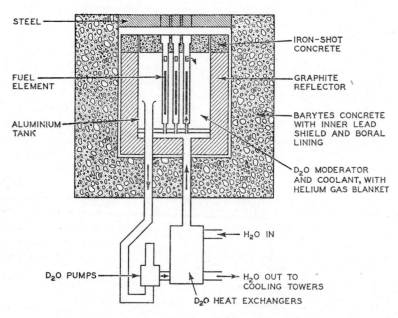

FIG. 15. THE DIDO RESEARCH REACTOR (HARWELL)

Thermal output	10 MW
Maximum thermal neutron flux	10^{14} n/cm²/sec
Fuel elements	U/Al alloy, clad in aluminium
D_2O investment	10 tons

radicals D and OD from gamma bombardment being sufficiently high to recombine all the molecular products formed from recoil deuterons obtained as a result of the neutron interactions with heavy water.

It is possible to upset this balance by the addition to the water of substances which will react preferentially with the radiation-produced free radicals and hence stop the recombination reactions. Chloride and bromide ions have been found to be extremely reactive towards H and OH radicals, for instance, and their presence in amounts of only a few parts per million can cause quite rapid net decomposition of irradiated water. Hence it has been found necessary to maintain the purity of water in a reactor at a very high level in order to achieve low decomposition rates (a specific conductivity of $\sim 1 \times 10^{-6}$ ohm^{-1}cm^{-1} is required, and is usually obtained by the application of ion-exchange processes). These considerations are not important for nuclear reactors cooled by ordinary *light* water, but they are vital to reactors employing *heavy* water since this is a very expensive commodity; if extensive decomposition occurred it would be necessary to install large external recombination equipment which would increase the complexity and cost of the reactor system.

So far we have discussed the chemistry of water used in research reactors such as NRX and DIDO, (Fig. 15) where the temperature is kept below boiling point and no attempt is made to extract useful heat. What happens in reactor systems where the temperature is raised to the region of 300°C in order to extract heat energy for electricity production? Provided that the water remains pure, it is just as resistant to radiation decomposition at high temperatures as it is at the lower temperatures, but a new problem emerges: that of the corrosion of the metals used to contain the circulating water. In the low temperature reactors the aqueous corrosion of structural materials such as aluminium and stainless steel is negligible. However, at 300°C it is not possible to rely upon aluminium as a constructional material in the reactor core, since the corrosion rate is high and the mechanical strength is low. Reactor designers have been forced to consider other metals of low neutron absorption cross-section for this application, zirconium being the favourite and the one on which most work has been done.

HIGH TEMPERATURE AQUEOUS CORROSION OF ZIRCONIUM

Constructional metals used in a nuclear power reactor core must be of low neutron absorption cross-section, otherwise the wastage of neutrons would be excessive and would increase significantly the cost of power produced.   Low cross-section metals are not very numerous and for zirconium to be acceptable for this use it is necessary to remove hafnium.   Normal zirconium minerals contain up to 2 per cent of the chemically very similar hafnium and the latter follows zirconium through all the normal concentration and purification stages.   Hafnium has the rather high neutron absorption cross-section of 115 barns and the presence of the naturally occurring amount is sufficient to raise the cross-section of pure zirconium, 0·18 barns, to over 1 barn.   Much chemical study has been devoted to zirconium—hafnium separations and several methods are now available.   For instance the two elements may be separated on a large scale by the use of anion exchange resins in hydrochloric—hydrofluoric, or sulphuric, acid solutions, or by preferential solvent extraction of the zirconium from a nitrate aqueous solution into tri-n-butyl phosphate diluted with an inert organic liquid such as kerosene.

Pure zirconium has been shown to have a very high degree of corrosion resistance to high temperature water, even if the water contains a wide variety of impurities.   However, the presence of certain impurities in the metal at quite low concentrations can have a disastrous effect on the corrosion rate.   The purest zirconium is obtained by refining the normal sponge metal (produced by reducing zirconium tetrachloride with magnesium) through the van Arkel—de Boer process.   The latter involves the formation of the volatile zirconium iodide from the crude zirconium and the subsequent decomposition of the iodide on a heated zirconium filament.   This process is obviously very costly to perform on a large scale and rigorous control of the conditions is required in order to achieve the desired level of subsequent corrosion resistance.   It is somewhat unfortunate that the impurity which has the most marked effect on corrosion resistance is nitrogen: this means that all operations on the metal during fabrication into reactor components must be performed in an inert nitrogen-free atmosphere of pure argon or helium.

Zirconium relies for its corrosion resistance upon the formation of a protective oxide film.   In the absence of impurities the corrosion rate is determined by the diffusion of oxygen ions from the water-oxide interface to the oxide-metal interface via vacant sites on the oxygen ion lattice in the film.   Consequently, as the oxide film builds up in thickness the corrosion rate falls.   The situation is complicated, however, by a change in kinetics when the protective film reaches a thickness of about 20 millionths of an inch; after this point the oxide film begins to flake off and protection against further corrosion is lost. The kinetics of the corrosion reactions follow the form:

$$\left.\begin{array}{l}\text{Weight oxide formed}\\\text{per unit area}\end{array}\right\} = K \text{ (time)}^n$$

where K and n are constants for a particular set of conditions. The point at which oxide begins to flake off the parent metal is marked by a change in the value of n from about one-third to unity: it is usually referred to as *break-away* (see Fig. 16).

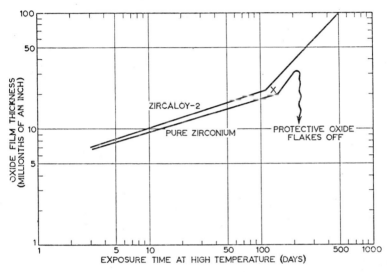

FIG. 16, BEHAVIOUR OF ZIRCONIUM AND ZIRCALOY-2 IN WATER AT HIGH TEMPERATURES

Experimental corrosion results are recorded in this way to show the build-up of the oxide film. The graph represents typical behaviour in water at 360°C. The exact position of the break-away point X depends upon the purity of the metal and upon the temperature.

The enhancement of corrosion by nitrogen impurity in the zirconium has been attributed to its incorporation in the growing oxide film as N^{3-} on oxygen ion sites. This increases the concentration of lattice vacancies, increases the oxygen ion diffusion rate, and causes the onset of break-away in a very short time. In order to achieve the desired degree of corrosion resistance for reactor application, it is necessary to keep the level of nitrogen impurity in the metal down to well below 100 parts per million. Another very harmful impurity in the zirconium is carbon. This means that in the manufacture of the metal it is not possible to perform melting and casting operations in graphite crucibles and the considerably more expensive arc-melting process with cooled copper hearths must be employed.

The exact cause of break-away has occupied the attentions of many investigators, particularly in the U.S.A., over the past ten years and several possible mechanisms have been suggested, none of which is entirely substantiated by all the evidence. Some of the most feasible mechanisms may be listed:

(a) Physical stresses may be built up in the oxide film due to the different lattice spacings of the oxide and the parent metal. These stresses may produce cracks in the oxide, thereby shortening the diffusion path for the oxygen ions and increasing the corrosion rate. Moreover, they would assist the spalling of the oxide.

(b) The overall corrosion reaction is essentially between zirconium metal and water molecules with the formation of oxide and hydrogen. Some of this hydrogen may diffuse through the oxide film to the metal-oxide interface, with the subsequent formation of zirconium hydride and loss of coherency between the metal and the protective oxide.

(c) The application of X-ray and electron diffraction techniques to the oxide films at the Battelle Memorial Institute showed that in the initial stages of corrosion a tetragonal form of zirconia was built up, and after break-away the oxide became monoclinic. The onset of break-away then appeared to be due to this phase transformation which may have been initiated by the growth of zirconium hydride inclusions in the underlying metal. These observations have not as yet been supported by any other laboratory.

A decision between these possibilities is not possible on present data and the elucidation of this problem remains a fascinating study for both the chemist and the metallurgist. The importance of the problem lies in the possibility of producing new alloys of zirconium with improved corrosion resistance; until the mechanism of corrosion is known, any decision on new types of alloy for investigation must remain essentially on an *ad hoc* basis.

An important development in the alloy field has been the introduction of " zircaloy-2 " based upon work in the Westinghouse Atomic Power Division Laboratories. The composition of this zirconium-base alloy is 1·5 per cent tin, 0·12 per cent iron, 0·1 per cent chromium and 0·05 per cent nickel. It is much cheaper to produce than *iodide* zirconium, being made directly from zirconium sponge metal. The addition of the tin has the effect of helping to counteract the disastrous effect of nitrogen impurity upon corrosion and perhaps of promoting the adherence of the protective oxide film to the metal. Although the initial corrosion rate in high-temperature water is somewhat higher than for the best quality iodide zirconium, the main advantage of this alloy is that the break-away point is much more reproducible than for the pure metal and the post break-away corrosion rate is lower (it is only 1×10^{-4} in. per year at 300°C). The effect of radiation upon this corrosion rate in a nuclear reactor is relatively small for systems operating on pure water, but it assumes serious proportions in the homogeneous aqueous reactors described later in this chapter.

STAINLESS STEEL CORROSION

Although zirconium or aluminium alloys, of low neutron absorption, must be used for the construction of the core components of a water-cooled reactor, there is no such restriction on the materials used in the non-irradiated parts of the reactor circuit. In fact, in a high temperature aqueous reactor the outer circuit must be made of steel in order to achieve sufficient mechanical strength combined with reasonable corrosion resistance. The austenitic types of chromium-nickel steels are generally considered to be satisfactory, but there are two aspects of the corrosion of these steels which have occupied considerable attention.

The first problem which has required investigation is the occasional cracking of stressed steel when in contact with high-temperature water. Considerable attention is being devoted in many laboratories to determine the exact mechanism by which this happens, since the problem has much wider implications than in the atomic energy industry alone. The problem appears to be worse in positions where there may be a water-gas phase interface, and it is certainly affected by very small traces of impurities (particularly chloride ion) in the water. Such considerations impose severe limitations on the impurity levels which can be tolerated in the water. Stress-corrosion cracking had a serious effect upon the Homogeneous Reactor Test (H.R.T.) constructed at Oak Ridge. Operation of the reactor was delayed about twelve months by traces of chloride ion having entered the leak detection system. There are many flanged joints on this reactor and each flange contains a specially built-in leak detector which is fed from a pure water reservoir (a leak produces a drop in pressure which can be measured). When the chloride-contaminated water came into contact with the stressed steel flanges at high temperature many small cracks appeared which, if left, would eventually have spread through the metal. It was found necessary to replace all of the flanges in the system in order to ensure that not one was left which might have been attacked in this way.

The other feature of steel corrosion under these conditions is the effect of traces of dissolved gases in the water. Small amounts of oxygen produce a considerably higher corrosion rate than that observed under reducing conditions; moreover, the surface oxide so produced may become loose and give rise to contamination of the water. Such contamination could lead to a higher rate of water decomposition in the reactor core and, furthermore, the particles would become radioactive by passage through the core and thereby increase the shielding requirements of the external circuit. Cases have also been observed when the steel corrosion products have been deposited on the zirconium cans of test fuel elements under irradiation, with a consequent serious effect upon heat transfer from the fuel to the water.

A solution to these problems has been found by maintaining a small excess of dissolved hydrogen in the water. Then stainless steel retains a bright appearance and there is very little formation of an oxide scale. The corrosion rate of ordinary mild steel is only about ten times greater than the austenitic stainless steels in high temperature water under reducing conditions. This corrosion rate would be acceptable provided that suitable methods could be found for continuously processing the reactor water to remove the corrosion products; ion exchange beds seem to be a convenient way of doing this since they would be capable of removing soluble impurities by exchange and insoluble ones by filtration through the bed. There is scope for the development of ion exchange materials which can be operated at high temperatures. If these problems can be overcome, mild steel may well have a wide application in this type of reactor, since it is cheap and relatively easy to fabricate.

HOMOGENEOUS AQUEOUS REACTORS

Solid fuel elements employed in many types of nuclear reactor have several serious disadvantages. Heavy costs are incurred in their fabrication to the close tolerances required, and in the use of exotic canning materials. Their working life is limited by structural and physical changes which occur under irradiation, and by the accumulation of fission product neutron poisons, and hence reprocessing must be done fairly frequently. The canning materials which are required to protect the fuel elements from chemical reaction with the reactor coolant have an adverse effect upon the neutron economy of the system. Finally, a large amount of expensive fissile material inevitably becomes locked up in the processing cycle, with a consequent increase in the overall cost of power generation. Homogeneous aqueous reactors seek to overcome these disadvantages by dispersing the nuclear fuel as a solution or suspension in water which acts as both moderator and coolant. Many reactors of this type are possible since we have a choice of two fissile elements (uranium and plutonium) and two fertile elements (uranium and thorium) which may be dispersed either separately or as mixtures in either light water or heavy water. The system which has been regarded in the Oak Ridge Laboratories, (U.S.A.), as showing the most promise of successful application consists of a core circuit containing a solution of uranyl sulphate (U^{233} isotope) in heavy

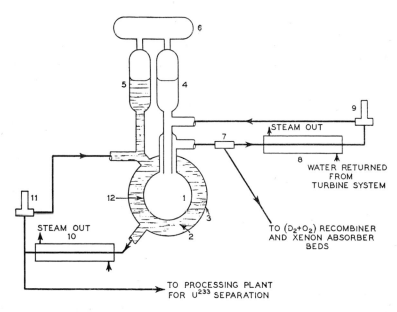

FIG. 17.   A Two-zone Homogeneous Aqueous Reactor, showing the Basic Circuits of One Possible Design of the Reactor

1. Core containing solution of uranyl sulphate
2. Blanket of aqueous thoria slurry
3. Pressure vessel, about 10 to 12 ft in diameter, to withstand 2000 p.s.i.
4. Core surge-tank
5. Blanket surge-tank
6. Gas-filled pressuriser

7. Radiolytic gas separator
8. Core-circuit heat exchanger (for a reactor of 100 MW electrical output nine heat exchangers might be required)
9. Core-circuit pump
10. Blanket heat-exchanger
11. Blanket circuit pump
12. Neutron " window "

water, surrounded by a "blanket" of heavy water containing a slurry of solid thorium dioxide particles. The basic circuits of this reactor are shown in Fig. 17, although the addition of the considerable number of auxiliary circuits required will make the whole plant much more complex. Reactor systems which employ suspensions of solid uranium oxide particles in heavy water are being studied by the Westinghouse Electric Corporation and in the Netherlands. The discussion which follows will be primarily about the chemical problems of the uranyl sulphate solution reactor, but a section is included on the special features of the suspension systems.

The homogeneous aqueous reactor is essentially a continuous chemical plant in which a highly exothermic reaction is taking place. It uses a feed of fertile material, produces heat for conversion to electricity with additional fissile material as a by-product, and fission products are eliminated as waste. Fuel can be added continuously to replace that which has undergone fission, and fission product poisons can be removed continuously while the reactor is operating. This removal of poisons, coupled with the absence of neutron absorbing materials normally required for covering fuel elements in heterogeneous reactors, leads to very favourable neutron economy. It appears to be possible to produce more U^{233} by capture of spare neutrons in the thorium blanket than is consumed in the reactor core by fission, so that after reaching equilibrium the reactor can be made to operate on a feed of thorium only and to produce excess U^{233} as a valuable feed material for other reactors. The design of such a reactor system produces a wealth of chemical problems.

CHOICE OF FISSILE SOLUTION

Uranyl sulphate solution has been chosen as the most desirable core fluid by a process of elimination. For a breeder reactor system the chloride, perchlorate, bromide, phosphate and chromate possess too high a neutron absorption cross-section. The choice then remains to be made between the fluoride, nitrate, carbonate and sulphate, all of which have quite sufficient solubilities in high temperature water for reactor use.

Dilute solutions of uranyl fluoride are not thermally stable above about 250°C, since hydrolysis of the uranium occurs with resultant precipitation. This difficulty can be overcome by the addition of sufficient hydrofluoric acid to the system, but a severe restriction on constructional materials is consequently imposed. Stainless steel has a low corrosion rate in this medium, but the metals such as zirconium and titanium which would form the basis of the neutron *window* between the reactor core and blanket are very severely attacked by even small traces of free fluoride ion. There appears to be no outstanding advantage of uranyl fluoride over the sulphate to counteract this disadvantage.

The use of a carbonate system would require a relatively high carbon dioxide pressure in the system to provide sufficient complexing of the uranium to maintain it in solution. Since aqueous reactors of this kind must operate near to the limit of engineering feasibility of the containing vessel due to steam pressure alone, the addition of a high overpressure of carbon dioxide is evidently undesirable.

Uranyl nitrate solutions have been used successfully in the low power research reactors HYPO and SUPO at Los Alamos, U.S.A. It is felt, however, that the use of nitrate in a high power, high flux, reactor would not be attractive owing to the instability of the nitrate ion under irradiation. The effect of fission fragments and of fast neutrons is sufficient gradually to reduce the nitrate down to nitrogen gas.

We are left with uranyl sulphate as the most promising salt for this reactor. The radiation stability of the sulphate ion is very good and the uranyl ion is also stable provided that oxidising conditions are maintained in the solution by the addition of a small overpressure of oxygen gas. At temperatures in the region of 300°C uranyl sulphate solutions are rather more stable thermally than was the case with the fluoride, but it is necessary to lower the pH to about 2 by the addition of excess sulphuric acid in order to prevent precipitation of the uranium. Moreover, the situation at high temperatures is also complicated by a region of two liquid phases in the phase diagram. If a particular uranyl sulphate solution is heated under pressure, at some definite temperature it will separate into two liquids: a dense phase containing most of the uranium sulphate, and a water phase. This would be very undesirable in a reactor, but fortunately at the uranium concentrations of interest for a power reactor (1–10 g/l.) the phase boundary is higher than the desired operating temperature, and is in fact raised by the addition of the sulphuric acid required to stabilise the solution against hydrolysis.

OTHER PROBLEMS OF THE FISSILE SOLUTION

Most of the circuit of a large homogeneous aqueous reactor would be constructed of stainless steel. Although the steel corrosion rate is comparatively low, rather less than 10^{-3} in. per year, the heat exchangers have a very large surface area and a considerable amount of metal undergoes the corrosion reaction. In a 400 MW reactor, the internal surface area would be about 25 000 ft^2 and it has been estimated that over one kilogram of stainless steel corrosion products would be produced for each day of operation. Iron from the steel would form insoluble ferric oxide, whereas the chromium constituent would first produce soluble chromate ions which would hydrolyse subsequently under the effect of the reactor radiations also to give an insoluble oxide. An auxilliary circuit has to be added to the main reactor circuit to remove these solids by centrifugal action in a hydroclone. The nickel from the steel corrosion presents a more serious problem, however, since it goes into solution as nickel sulphate. If this is allowed to build up in concentration, there is the possibility of loss of uranium fuel from the solution by the formation of insoluble precipitates. To prevent this happening, it is necessary to ensure that the nickel concentration remains below about 0·02 M, and this entails the addition of a further auxiliary processing circuit capable of treating up to 1 per cent per day of the reactor solution.

In the early considerations of this type of reactor system it appeared that a major difficulty might arise due to the rapid decomposition of water by the uranium fission recoil particles. It was shown on page 117 that water may decompose by two different routes according to which type of radiation is present. Fission fragments are intensely ionising along their tracks and hence they favour the molecular decomposition reaction with the formation of deuterium and deuterium peroxide. At the reactor operating temperature of 300°C, the peroxide is decomposed rapidly by the normal thermal process and the overall result is the production of a mixture of deuterium and oxygen. The magnitude of the problem may be judged from the estimate that a 400 MW (heat) reactor would give rise to 30 lb/min of deuterium and 120 lb/min of oxygen. Besides presenting a considerable explosion hazard, the recombination of this mixture to heavy water and subsequent return of the latter to the main reactor system would have been a major chemical engineering problem. In the studies on homogeneous aqueous systems at Oak Ridge, U.S.A., an attempt was made to overcome this gas evolution problem by the development of a catalyst which could promote the recombination in the reactor core. Fortunately, the investigations were successful, and it was observed that cupric sulphate had sufficient catalytic activity for reactor use. Its efficacy was proved by extension of the laboratory scale investigations to experiments in the 1 MW Homogeneous Reactor Experiment (H.R.E.). The rate-determining step of the catalytic reaction appears to be between copper ions and hydrogen (or deuterium) *in solution* and may be written in either of two ways:

$$Cu^{++} + H_2 \rightarrow CuH^+ + H^+$$
$$or \ Cu^{++} + H_2 \rightarrow Cu^+ + H_2^+$$

Present evidence is not sufficient to make an unambiguous choice between these two possibilities; in either case the next step is a rapid reaction between the reduced copper species and dissolved oxygen with the regeneration of cupric ions. The kinetics of the rate-determining recombination step have been shown to be of the form:

$$-\frac{d[H_2]}{dt} = k[H_2] \ [Cu^{++}]$$

Thus a particular concentration of copper sulphate will promote a fixed rate of recombination in the reactor and, if this is less than the rate of production of the deuterium-oxygen mixture by radiolysis of the water, an increase in pressure will occur. In practice it is expected that sufficient copper salt would be added to make the recombination rate about 90 per cent of the production rate, the excess 10 per cent being bled off the system continuously in order to sweep out the fission product neutron poisons, xenon and krypton. The gas mixture is then diluted with steam until it is below the known explosion limit and is passed through a catalytic bed of platinum supported on the surface of alumina for recombination. The steam is condensed and the radioactive rare gases are adsorbed on charcoal for storage and disposal.

THE NEUTRON WINDOW

One of the major obstacles to the present-day construction of highly rated power reactors of the two-zone type is the question of the constructional material for the core vessel which separates the fissile core solution from the surrounding fertile blanket. This material must be transparent to neutrons (hence the name *window*) and must have reasonably good mechanical properties at 300°C in order to withstand the effects of accidental misbalance of the pressures in the core and blanket circuits. Aluminium would not meet the latter requirement and little is known of the properties of low cross-section beryllium. The reactor designer is forced to consider hafnium-free zirconium as a basis for the window. Although this metal, and its alloy zircaloy-2, have been found very suitable for use in heterogeneous water-cooled reactors, under the conditions in the homogeneous reactor the corrosion rate would be completely unacceptable. The corrosion rate increases with the rate of fission in the solution with which it is in contact, but it is not clear on present evidence why this should be so.

Several possible ways have been suggested by which the rate of fission in solution could influence the zirconium corrosion process:

(*a*) Radiation may produce short-lived chemically reactive species in solution which would perhaps increase the dissolution rate of the protective oxide film.

(*b*) Zirconia is a very poor conductor of heat; as the protective oxide film builds up, thermal gradients may be introduced which could affect diffusion conditions at the metal-oxide interface or within the oxide itself. The effect would be increased under homogeneous reactor conditions owing to the great increase in energy deposited in the zirconium oxide film from the fission fragments not present under heterogeneous reactor conditions.

(*c*) Energy deposition in the protective oxide film on the zirconium may increase the concentration of defects in the crystal lattice. Since the corrosion appears to take place by means of diffusion of oxygen through vacant oxygen ion sites, an increase in the vacant site concentration may well increase the corrosion rate.

(*d*) Energy deposition in the protective zirconia film may cause a transformation of the usual monoclinic form of the oxide to the cubic form which is normally stable only at very high temperatures. Physical stresses would thus be set up within the oxide film owing to the volume change involved in the phase transformation, with consequent spalling of the outer layers of oxide. There is some evidence that such a transformation takes place upon fission fragment bombardment of zirconia, but doubt remains as to whether the energy deposition rate under reactor conditions would be high enough to cause the transformation to occur rapidly enough to affect the corrosion rate.

The complexity of the problem is illustrated by the following table, which gives the predicted rates of energy deposition from various sources in a zirconium film 10^{-4}cm thick under power reactor conditions:

Radiation	*Energy Deposition Rate* ($MeV. cm^{-2} sec^{-1}$)
Fast neutrons	$1\cdot4 \times 10^8$
Beta decay energy	3×10^9
Instantaneous gamma energy . . .	$1\cdot4 \times 10^{10}$
Fission recoil fragments	$2\cdot3 \times 10^9$

Thus a major part of the energy deposited in the protective oxide film arises from the beta and gamma fluxes within the reactor. However, the effect of the various radiations is likely to be quite different: an estimate of the rate of production of defects in the oxide lattice shows that fission fragments are about five million times more effective than fast neutrons and ten million times more effective than fast beta particles. These differences in turn will be somewhat evened out by several annealing processes which can cause the disappearance of the defects.

In order to understand the mechanism of corrosion it will be necessary to attempt to separate all these different effects and to study each one separately. Only then will it be possible to predict what type of zirconium alloy is likely to have the desired corrosion resistance. If a successful approach along these lines is not feasible, there are several physical methods which may be considered for controlling the corrosion.

It may be possible to clad the inner surface of a zirconium alloy core vessel with a thin layer of another metal of higher corrosion resistance but higher neutron cross-section. For instance, gold has been used as a cladding material in the LAPRE I and II reactors (Los Alamos, U.S.A.) for corrosion resistance against concentrated phosphoric acid core fluid. However, for a large power reactor the choice would have to be another noble metal rather than gold since the high neutron flux would convert the latter rapidly into mercury!

A second physical method of controlling the corrosion might be to apply a thick impervious coat of oxide to the metal. Considerable progress has been made over recent years in the techniques of flame spraying coatings of zirconia and alumina on to metals, but there is no known method at present for making the coating completely impervious—there is always a small number of open pores through which solution could reach the metal and begin the corrosion process.

Finally, if it can be shown that the damaging effects are due to fission fragments rather than to the other radiations, then it has been suggested that the core vessel could be made with a porous inner lining through which pure water could be fed in order to maintain a pure water film about one thousandth of an inch thick between the metal and the core fluid.

All the fission fragment energy would then be absorbed in the water film rather than in the metallic oxide. The technology of the construction of such a system would obviously be difficult.

OXIDE SUSPENSIONS

Even if the corrosion problem in the aqueous solution system proves to be a major obstacle to future development of large reactors of this type, it may still be possible to retain the advantages of the homogeneous system by utilising a suspension of a uranium compound as the core fluid. An oxide would be the first choice for such a suspension, and two types of reactor system may be feasible: an oxidised system based on uranium trioxide and a reduced system based on uranium dioxide.

A considerable amount of effort has been expended in the U.S.A. to elucidate the chemistry of the oxidised system and similar work on a smaller scale has been done at Harwell. Uranium trioxide is stable in high-temperature water, but it forms several hydrates which have different crystal structures:

Hydrate	*In Contact with Water*
$UO_3,2H_2O$	Stable at low temperatures
$UO_3,0\cdot8H_2O$	Stable up to 180°C
	Long, thin rods
UO_3,H_2O	Stable between 200 to 280°C
	Flat plates
$UO_3,0\cdot5H_2O$	Stable above 280°C

It is unfortunate that one of these phase changes occurs in the temperature region of interest for reactor operation, so that prediction of the equilibrium state is rather difficult. Hence the characteristics of the aqueous suspension of importance for determining the ease with which the latter may be handled also become difficult to assess, and they may change with time. A further trouble could arise through the fact that an aqueous suspension of enriched uranium (or of U^{233}) would be quite dilute, perhaps up to 10 g/l. and an appreciable fraction of this could be lost by adsorption on the enormous surface area of the heat exchangers.

The reduced system based on a suspension of uranium dioxide is under study in the Netherlands and is the basis of the design for the reactor SUSPOP. An atmosphere of hydrogen is necessary to prevent oxidation of the uranium, but otherwise the system is simpler than the oxidised one since no hydrates are formed. Interesting chemical problems arise from this system since the proposed method of operation is to maintain the uranium oxide particle diameter below 13μ so that a large fraction of the fission fragments will be ejected from the particles into the aqueous phase. The rare gases will then be easily removable, some of the fission products will remain in solution and some will readsorb on to the oxide surface. Small particle size charcoal added to the system in amounts of about 2 g/kg oxide, will preferentially adsorb fission products of the lanthanide types. It is believed that this will enable continuous removal of poison to be achieved by separation of the charcoal from the uranium oxide in an external by-pass circuit.

Thorium oxide suspensions are of vital interest to the two-zone reactor system. The fertile blanket which forms the outer zone of the reactor may contain either U^{238} to breed plutonium or Th^{232} to breed U^{233}; the latter is the more favoured since U^{233} is a valuable fissile material, whereas plutonium will be formed in large quantities as a by-product from the natural uranium gas-cooled reactors which are the basis of the present reactor construction programme in Britain. However, to make the system breed more U^{233} in the blanket than is simultaneously used up by fission in the core, it is necessary to make the thorium very concentrated, at least 1000 g per litre.

The only thorium compound which is known to have a solubility in this region is the nitrate, but its use is ruled out by radiation decomposition of the anion. The nitrate would be reduced to nitrogen at a rapid rate and it would be necessary to design an external circuit to reform the nitrate and return it to the reactor. It would not be permissible to allow any of this nitrogen to go to waste since the effect on the neutron economy of large quantities of nitrate imply that only low cross-section, expensive, separated N^{15} could be used. Natural nitrogen would absorb too many neutrons which should otherwise be put to good use by absorption in the thorium.

In the absence of a suitable thorium solution it is necessary to consider the use of a concentrated suspension of the oxide. Little general information of direct application to the technology of thoria suspensions was available before the concept of the homogeneous aqueous reactor, and considerable effort is being devoted to this subject both at Harwell and in the United States. Suspensions of this kind can exist in a flocculated or a dispersed (deflocculated) state and it is important to investigate the factors affecting this since the particular state has a direct bearing on the problems of handling. A dispersed slurry behaves in a similar manner to a normal Newtonian fluid, whereas a flocculated slurry behaves as a Bingham plastic (it has a definite yield stress and its coefficient of rigidity is much higher than the viscosity of the corresponding Newtonian fluid).

The disadvantage of a flocculated slurry is that it may settle to form hard cakes of solid which are difficult to redisperse. Factors which can affect the flocculation of the suspension are temperature, impurities such as the accumulation of stainless steel corrosion products, substances which are deliberately added to change the state, and perhaps radiation.

Information concerning the interrelation of these variables is obtainable by a study of electrokinetic phenomena which occur at the oxide-water interface. The surface electrical charge on the thoria appears to be a factor of importance in determining the behaviour of the aqueous suspension. This charge originates from ionisation of surface hydroxyl groups produced by hydration of the thoria surface:

$$\mathrm{ThO_2 + H_2O} \rightarrow \quad \mathrm{Th} \begin{cases} \mathrm{OH} \\ \mathrm{OH} \end{cases} \rightleftharpoons \quad \mathrm{Th_+^+} \begin{cases} \mathrm{OH^-} \\ \mathrm{OH^-} \end{cases}$$

The hydroxyl ions then reside in the diffuse electrical layer in the aqueous phase. The dotted line represents the boundary between the compact and diffuse electrical layers. The surface hydroxyl groups can also ionise in a different way to produce a surface charge of opposite sign:

$$\mathrm{Th} \begin{cases} \mathrm{OH} \\ \mathrm{OH} \end{cases} \rightleftharpoons \quad \mathrm{Th} \begin{cases} \mathrm{O^-} \ \mathrm{H^+} \\ \mathrm{O^-} \ \mathrm{H^+} \end{cases}$$

There is some evidence that irradiation can cause a shift from the original predominantly basic surface dissociation to a more acidic surface.

Phenomena connected with the surface electrical properties are usually investigated by observations of electrophoresis. The application of such techniques to suspensions at high temperature and pressure, and ultimately under irradiation, presents a stimulating problem for the future.

PYROPHORICITY

The use of zirconium in nuclear reactors has led to a very rapid expansion in the tonnage of metal manufactured and fabricated by industry. The large-scale handling of a relatively new metal has not been achieved without hazard: despite precautions taken with finely divided metal which was known to be in a highly reactive state, serious accidents involving fires and explosions have occurred. Similar problems and hazards have been met in the large-scale handling of thorium and uranium.

Some metals (e.g. aluminium) are theoretically capable of reacting with water to produce about the same explosive energy as T.N.T. Moisture certainly seems to play a significant part in the initiation of pyrophoric reactions by the production of heat and hydrogen. The latter can then react to form the metal hydride which is usually very reactive. Stress, such as that produced in lathe-turnings, is also believed to have a bearing on the initiation of explosive pyrophoric reactions. Although most of the accidents of this type have occurred with finely divided metal (which has a large surface area available for reaction and a small percentage of surface area exposed directly to the surroundings for loss of heat by radiation), cases have occurred also when similar reactions have taken place with bulk metal. The detailed chemical and metallurgical features of these pyrophoric reactions are far from clear at the present time.

FURTHER READING

Charpie, Hughes, Littler and Trocheris. *Progress in Nuclear Energy*, Series II. *Reactors,* Vol. 1. Pergamon Press, 1956.
United Nations. Proceedings of the First International Conference on the Peaceful Uses of Atomic Energy, Geneva, 1955. Vol. 9, *Reactor Technology and Chemical Processing.*

CHAPTER 8

GAS-COOLED REACTOR SYSTEMS

THE first reactors to be built in the United Kingdom were graphite-moderated and air-cooled, with fuel elements of natural uranium metal canned in aluminium. The choice was influenced by many considerations, including the nuclear stability of this type of reactor and the comparative cheapness and good machining properties of graphite. The experience gained in the construction and operation of the GLEEP (1947), BEPO (1948) and WINDSCALE (1950–51) reactors suggested that gas-cooled reactors could be developed to produce useful power. Many new problems had to be overcome, however; these arose mainly from the higher temperatures and neutron fluxes which obtain in a power reactor. It is no small step from a research reactor such as BEPO to the prototype power reactor at Calder Hall, as shown by the data presented in Table 18.

TABLE 18.—COMPARISON OF THE OPERATING
CONDITIONS OF THE BEPO AND CALDER HALL REACTORS

	BEPO	Calder Hall
Heat output (megawatts)	6·5	180
Max. thermal neutron flux (neutrons $cm^{-2}sec^{-1}$)	$\sim 2.10^{12}$	$\sim 2.10^{13}$
Coolant outlet temperature (°C)	90	336
Max. can temperature (°C)	200	408
Average rating (MW per ton of uranium)	0·15	1·4

The BEPO and Calder Hall reactors are depicted schematically in Fig. 18.

PROBLEMS IN THE DEVELOPMENT OF THE CALDER HALL REACTORS

THE GRAPHITE MODERATOR

For many years graphite has been produced artificially and widely used in industry. Its general properties were therefore well-known at the time when the reactor designs were begun. It was found,

136

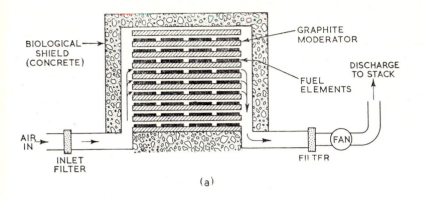

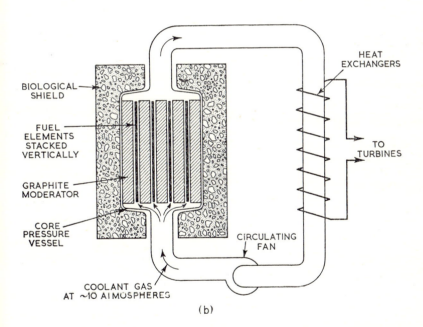

FIG. 18. GAS-COOLED REACTORS

(a) Air-cooled research reactor (e.g. BEPO).
(b) Carbon dioxide cooled power-producing reactor (e.g. Calder Hall).

however, that under irradiation an important new phenomenon occurs. This is known as the *Wigner effect*, and arises from damage of the graphite structure during irradiation.

Radiation Damage of Graphite

During the process of slowing down, the fast neutrons resulting from fission transfer their energy to the atoms of the moderator by collision. Most of this energy appears as heat, but some of it causes displacement of carbon atoms from their normal positions in the graphite lattice. Although these interstitial atoms are at a higher potential energy than atoms in the crystal lattice, some energy of activation is required before they can return to their original positions. At room temperature the energy of thermal agitation is not sufficient to accomplish this at a significant rate, but as the temperature is raised an increasing proportion of the displaced atoms are able to return to their normal lattice sites. The presence of displaced atoms in the lattice gives rise to several observable changes in the properties of the graphite. These changes are not only of interest as a means of studying the nature and extent of the radiation damage, but are also of vital importance to the reactor designer. Some of the changes which have been observed on samples of graphite after irradiation are listed below.

(a) *Crystal Structure.* The displacement of carbon atoms from their positions in the lattice results in a distortion of the structure of the graphite crystallites. This is illustrated by the X-ray diffraction-patterns of irradiated samples; the diffraction lines show marked changes in position and considerable broadening. From the shifts of the diffraction lines it is inferred that the interplanar (c_0) spacing is increasing, and the average distance between the atoms within the planes is decreasing, with increasing exposure. During a typical prolonged irradiation the average c_0 spacing increases from 6·7Å to 7·5Å.

(b) *Dimensional Changes.* As may be anticipated from the observed changes in the lattice parameters, changes in the dimensions of the graphite occur. It is impossible to generalise on the effects observed, because these depend upon the nature of the original graphite sample. The production of graphite normally involves either an extrusion or a moulding process, both of which impart some degree of orientation of the graphite crystallites in the formed mass. As the change in lattice spacing of these crystallites on irradiation is not isotropic, the dimensional changes of a block of graphite are different along the different directions. The increase in length observed at the highest irradiation exposures usually amounts to about 5 per cent for artificial graphite, but an increase of 24 per cent has been observed on a sample of natural graphite.

(c) *Thermal and Electrical Conductivity.* The thermal conductivity shows a steady decrease with increasing exposure, falling to about one-fiftieth of the unirradiated value at the highest exposures studied. The electrical conductivity, on the other hand, very quickly drops to about one-fifth and then remains practically constant. Furthermore, the change in electrical conductivity is less sensitive to the irradiation temperature than is the thermal conductivity.

(d) *Heat Content.* The presence of atoms of carbon occupying interstitial positions in the normal graphite lattice, with subsequent distortion of the lattice, implies that the energy content of the graphite is increased, the increase being known as the *stored energy.* This may be measured either by comparing the heats of combustion of graphite samples observed before and after irradiation or by direct measurement, when the stored energy is released by heating the sample in a calorimeter. The amount of stored energy can be quite considerable, amounting to 500 cal/g in prolonged irradiations. This represents sufficient heat to raise the temperature of the sample some 1200°C under adiabatic conditions.

The exact mechanisms of the radiation damage of graphite are not yet fully understood, because of the large number of variables involved and because several different mechanisms appear to be responsible for the observed effects. Several general conclusions can be drawn, however. In irradiations carried out at room temperature the extent of damage increases with the duration of exposure, and usually shows no indication of reaching a saturation condition, even after prolonged irradiation. At higher temperatures the damage builds up more slowly; for example, after the same radiation dose the extent of damage in samples irradiated at 100°C and 200°C respectively is only 35 per cent and 15 per cent of the damage at 30°C. At still higher temperatures the rate of damage is even less, and tends to reach saturation.

Annealing of the Irradiation Damage

By heating the graphite to some suitable temperature the irradiation damage can be annealed out. Like the mechanism of damage, the mechanism of annealing is complex and not fully understood. The process of annealing begins to take place at temperatures some 50 to 100°C above the irradiation temperature and at successively higher temperatures a greater degree of annealing occurs. The ease with which graphite can be annealed (as measured, for example, by the temperature to which the sample must be heated in order to release a given proportion of the stored energy) depends upon the temperature of irradiation and upon the total irradiation dose. In irradiations carried out at room temperature most of the stored energy is released on heating to about 250°C after short irradiations. As the dose is increased annealing becomes more difficult and higher temperatures are required. After very prolonged irradiation annealing is not fully complete until a temperature of 2000°C is reached.

Effect of Radiation Damage on the Operation of a Nuclear Reactor

The two most important consequences of radiation damage are the physical growth of the graphite and the presence of stored energy. In a low-temperature reactor such as BEPO it is necessary to anneal the graphite moderator periodically by suitable heating. This restores the graphite to its original dimensions and prevents the build-up of stored energy to such proportions that its inadvertent release would present a considerable hazard. In a power reactor, with its higher operating temperature, these proplems are less acute, for the rate of damage is much reduced and self-annealing tends to take place.

THE COOLANT GAS

Air is used as the coolant for the reactors operating at comparatively low temperatures without imposing any risk of causing excessive oxidation of the graphite. At the higher temperatures obtaining in a power reactor the rate of oxidation would become excessively high, and a coolant more inert than air is therefore required. Besides chemical compatibility, other factors which limit the choice of coolant gas are availability, heat-transfer properties, radiation stability and neutron capture cross-section.

Of the possible gases, helium is expensive, hydrogen would diffuse through the canning material and react with the hot uranium to form hydrides, nitrogen has a high neutron absorption and organic vapours, such as benzene, are unstable in the presence of radiation. Carbon dioxide was chosen therefore for the coolant in the Calder Hall and subsequent reactors. Two important questions required an answer before the final decision could be made. Firstly, to what extent does carbon dioxide decompose under pile irradiation? Insertion into BEPO of sealed silica tubes containing this gas showed that negligible decomposition to carbon or carbon monoxide and oxygen took place. Secondly, how is the equilibrium between carbon dioxide and graphite disturbed by the presence of radiation? In the temperature range of interest (300 to 400°C) the thermal equilibrium for the reaction

$$C + CO_2 \rightleftharpoons 2CO \qquad (1)$$

lies well over in favour of carbon dioxide, as shown by the values in Table 19.

TABLE 19.—COMPOSITION OF THE GAS RESULTING FROM THE REACTION OF CO_2 WITH GRAPHITE (% CO BY VOLUME)

Temperature	300	400	500	600	°C
CO, pressure = 1 atmos.	0·006	1·0	7·2	36	% CO
CO_2 pressure = 10 atmos.	0·002	0·3	2·3	11	% CO

If carbon monoxide were produced in appreciable amounts within the reactor core, and it decomposed again at the lower temperatures in the heat exchangers, two very serious consequences could result. Loss of graphite from the moderator might appreciably shorten the useful life of the core, and any deposition of carbon in the heat exchangers would impair their efficiency.

From a knowledge of the operating conditions in a power reactor it was possible to set an approximate upper limit to the rate at which carbon dioxide could be permitted to react with the graphite. This turned out to be greater than the reaction rate in the absence of radiation. Nothing was known, however, about the influence of radiation on the reaction and so experiments were designed to determine the magnitude of this effect. Samples of graphite were irradiated in a furnace in BEPO and a stream of pure carbon dioxide was circulated over them in an apparatus constructed entirely of glass and silica. As the anticipated loss in weight of the graphite sample was only of the order of micrograms, the extent of reaction was observed by determining the carbon monoxide content of the gas stream. The amounts observed were, in fact, greater than those arising from a purely thermal reaction, but the increase was still below the maximum set by the design engineers.

By varying the gas flow-rate and the shape and size of the irradiation vessel it was shown that the radiation-induced reaction was initiated by the formation of chemically reactive species in the gas-phase, rather than by activation of the graphite itself. It follows that the observed rate of reaction would depend not only upon the rate of production of these active species in the gas phase—and hence upon the radiation intensity—but also upon the exact flow-pattern within the system. This determines the rate at which activated species reach the graphite surface. A full scale reproduction of a power reactor coolant channel was therefore constructed in BEPO, with a view to simulating as closely as possible the conditions likely to be met in the power reactor. It was finally confirmed that transport of graphite by this mechanism would not constitute a serious problem at the power densities envisaged in the Calder Hall reactors.

Gas Purity

The presence of water vapour in the carbon dioxide can give rise to several important side reactions following on reaction 1, such as:

$$H_2O + CO \rightleftharpoons CO_2 + H_2 \qquad (2)$$

$$2U + 3H_2 \rightleftharpoons 2UH_3 \qquad \text{(H}_2 \text{ diffuses through the fuel element can)} \qquad (3)$$

$$4Zr + 3H_2 \rightleftharpoons 4ZrH_{1.5} \qquad \text{(Zr is a possible canning material)} \qquad (4)$$

The thermodynamic data for these reactions are well known, although with the exception of reaction 1 the influence of radiation is not established. It is unlikely that a true equilibrium will be attained at any point in a flowing, non-isothermal circuit; the actual conditions obtaining will depend upon reaction rates and the time of transit of the gases through the relevant part of the circuit. Qualitative conclusions can be drawn, however, from a consideration of the calculated thermodynamic equilibrium conditions. Removal of water vapour by reaction 2 could give rise to difficulties in drying the gas after it has been introduced into the main reactor circuit, and appreciable amounts of hydrogen would be produced by this reaction. These amounts would not, however, be sufficient to produce uranium hydride by reaction 3, but could lead to breakdown of any zirconium present by reaction 4. Conversely, the reverse of reactions 2 and 1 would maintain any hydrogen impurity at the equilibrium value should hydrogen be introduced into the system by, for example, chemical break-down of the lubricants in the circulating fans.

THE FUEL ELEMENTS

In the BEPO and Windscale reactors the fuel elements were canned in aluminium in order to prevent oxidation of the uranium and also to retain the fission-products. At the higher temperatures employed in power reactors aluminium is no longer suitable.

Magnesium is a possible alternative, for it does not react with uranium and has a low neutron absorption.   The use of the pure metal as a canning material, especially in air-cooled reactors, introduces a serious fire risk, and is limited also by undesirable metallurgical properties at elevated temperatures.   Magnesium alloys, known as *Magnox*, of suitable mechanical properties and more resistant to oxidation either by air or carbon dioxide, have been developed and are employed in the Calder Hall reactors. These alloys promise to be suitable for can temperatures of up to 450°C, corresponding to gas outlet temperatures of about 380°C.

Uranium metal itself undergoes physical changes under irradiation, the anisotropic crystals of uranium becoming elongated in one direction.   In a fuel element, which consists of randomly-oriented crystals, severe wrinkling of the surface results from this anistropic growth, with subsequent rupture of the can (See Plate VIII).   The deformation can be minimised by suitable heat-treatment of the uranium prior to irradiation.   This involves quenching the hot fuel element to give a random fine-grain structure which is less susceptible to wrinkling.

FUTURE DEVELOPMENTS

The gas-cooled reactors at Calder Hall have sometimes been referred to as the " Model-T Ford " of reactors, but, just as the

TABLE 20.—POSSIBLE DEVELOPMENTS IN GAS-COOLED REACTORS
(Approximate gas temperatures in parentheses)

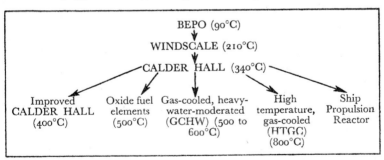

Model-T was the forerunner of more advanced and more sophisticated models, so the prototype reactors at Calder Hall are capable of much improvement and development.   Some possible lines are indicated in Table 20.

The object in all these developments is to achieve:

  (i) A higher working temperature, giving increased efficiency in the generation of electricity;

 (ii) a higher output per reactor, so reducing the capital cost per kilowatt generated; and

(iii) a longer life for the fuel element, to reduce the cost of processing and refabrication.

The Ship Propulsion Reactor is an exception, for here the object is to cut down the size and weight of the system by the use of enriched uranium. As these are problems of engineering design rather than of a fundamental nature this reactor will not be discussed further.

IMPROVED CALDER HALL

The reactors now being planned and built by the various consortia (at Bradwell, Berkeley, Hunterston and Hinkley Point) are all basically of the same form as the Calder Hall reactors, but striking improvements in performance have been accomplished by refinements in the engineering design. For example, in the Hinkley Point reactors the use of a spherical core pressure vessel of 3-in. thick plate, compared with the 2-in. thick cylindrical core vessel at Calder Hall, permits an increase in coolant gas pressure from 115 p.s.i. to 180 p.s.i. This increase, together with other improvements, will result in an electrical output per reactor of 250 MW, compared with the 48 MW at Calder Hall. No chemical problems, however, arise with such refinements of design.

The temperatures attainable within the core of these improved reactors are still limited by the maximum of about 450°C imposed upon the fuel element surface temperature by the Magnox canning alloy. Even if this were replaced by another metal of higher melting-point, such as beryllium, the properties of the uranium metal itself would impose a further limitation on the temperature. With a can temperature of 450°C the temperature of the centre of a typical cylindrical fuel element is of the order of 600°C. Any attempt to improve the performance of the reactor by operating at a higher power density, and hence with a greater heat output per fuel element, would result in even higher uranium temperatures. Unfortunately uranium metal undergoes phase transformations at 660°C and 770°C, and at the upper temperature a considerable change in volume occurs. At the higher heat ratings the maximum uranium temperature could be maintained below the transition-point by a suitable change in shape of the fuel element, from a cylinder, to, for example, a thin plate. This would necessitate the introduction of a greater amount of canning material into the reactor, with a subsequent increase in neutron capture. To offset this, some enrichment of the fuel would be required. The cost of enrichment, coupled with the increased cost of a novel canning alloy, would probably outweigh the advantages gained by the increased ratings and temperature.

OXIDE FUEL ELEMENTS

The refractory nature of uranium dioxide would seem, at first sight, to make this an ideal alternative to uranium metal as a fuel element for operation at higher temperatures. There are, however, many problems to be solved before oxide fuel elements can be used with confidence in highly rated reactors. Compacted uranium dioxide is known to show severe cracking when subjected to prolonged reactor irradiation and this reduces the effective thermal conductivity, which in any case is lower than the conductivity of uranium metal. Moreover, to achieve good heat conduction away from the oxide it is necessary to have a very closely fitting metal can. If the oxide shows any tendency to expand under irradiation it will then be necessary to strike a balance between a can thick enough to restrain physically the oxide and the increased loss of neutrons which a thicker can implies. As an alternative to a tightly fitting can the heat could be transferred by filling any gaps between oxide and can with a metal which is liquid at the operating temperature. A dead-space provided at the end of the can would accommodate any expansion in the oxide. Such a scheme immediately raises problems of compatibility of the liquid metal on the one hand and the oxide and can metal on the other hand.

For the can, metals of suitably low neutron absorption and high melting-point are zirconium (m.p. 1860°C) and beryllium (m.p. 1300°C). Zirconium reacts with carbon dioxide at temperatures above 500°C, but beryllium has no reaction up to temperatures of 600°C. Much effort, however, is required before beryllium can be produced of sufficient purity and in sufficient quantity for use in nuclear reactors. Such development, and the subsequent use of the metal, will be considerably hampered by the stringent safety precautions necessary when handling the highly toxic beryllium compounds.

THE GAS-COOLED HEAVY-WATER-MODERATED REACTOR (G.C.H.W.)

At the higher gas temperatures obtainable with beryllium-canned fuel elements, compatibility problems of graphite and carbon dioxide may arise, as indicated by the values of carbon monoxide content in Table 19. Furthermore, when using a graphite moderator it will not be possible to take full advantage of the higher ratings which are obtainable with oxide fuel elements because of the enhanced reaction of the graphite with carbon dioxide in the higher neutron flux. Such difficulties can be circumvented by the use of heavy water as the moderator. In this case it is also possible to enrich the fuel with plutonium produced as a by-product in the irradiation of natural uranium; this procedure is not altogether satisfactory with a graphite-moderated reactor because, for nuclear reasons, instabilities could arise in such a system.

The reactor is envisaged to consist of a tank of heavy water pierced by an array of fuel-element channels. These consist of two concentric tubes, the small gap between them being used for thermal insulation of the moderator. The carbon dioxide coolant flows over the fuel elements which are located along the axis of the inner tube. In a 500 MW (electrical) reactor it is estimated that 110 MW of heat will be generated within the moderator, largely by deposition of energy from fast neutrons and gamma radiation. The heavy water is therefore cooled by circulation through external heat exchangers, the temperature being maintained below 100°C so that no steam pressure is built up in the moderator circuit.

In such a reactor concept most of the chemical problems which do arise have either been met in other gas-cooled systems or have been solved in other connections. For example, radiolytic decomposition of the heavy water will occur, but the difficulty of gas recombination has already been overcome in the DIDO and NRX heavy water research reactors. At the low moderator temperature envisaged serious corrosion problems are not anticipated.

One interesting possibility is presented by the presence of a fluid moderator; that is control of the power level of the reactor by dissolving neutron-absorbing salts in the heavy water. This may be done instead of, or in conjunction with, the more conventional absorbing control rods. Of the dozen or so elements with sufficiently high capture cross-section, all, with the exception of boron can be eliminated on the grounds of either low solubility or high cost. Boric acid would be the preferred compound because of the absence of any induced activity, such as would result from the use of sodium borate. The main chemical problem arises in the removal of the boric acid when an increase in reactor power is required. Possible methods are precipitation followed by removal in a hydroclone separator, or absorption on an ion-exchange resin.

Although elegant in principle this method of control may in fact be difficult to apply in practice simply because of the high rates of injection and removal needed for the efficient and safe operation of the reactor. One important factor governing the size, and hence the cost, of an ion-exchange system is the rate and extent of uptake of the $H_2BO_3^-$ ion by the resin. The situation is complicated by the possibility of contamination of the heavy water by the carbon dioxide from the coolant system. Under unfavourable circumstances this could completely saturate the ion-exchange column in preference to the boric acid. A further complication arises by the production of Li^7 from the B^{10} (which occurs to an extent of 18·8 per cent in natural boron) by the (n,α) reaction

$$B^{10} + n \rightarrow Li^7 + \alpha$$

In a power reactor Li^7 would build up at a significant rate, so altering the pH of the moderator. Some means of removing the lithium, for example cation exchange, would therefore be required.

IMPREGNATED GRAPHITE FUEL ELEMENTS (H.T.G.C.)

As an alternative to using uranium oxide as a means of avoiding the difficulties arising from the phase transformations occurring in uranium metal at elevated temperatures, the uranium could be dispersed in an inert refractory material such as graphite. The refractory nature of the graphite would permit the attainment of fuel-element temperatures in excess of 1000°C, with gas outlet temperatures of the order of 800°C. With such temperatures the coolant gas could be used directly as the working fluid in a gas turbine, so omitting the inefficiencies which arise in steam generation.

Such a reactor concept has been developed by the U.K.A.E.A. under the name of the High-Temperature Gas Cooled Reactor (H.T.G.C.). The fuel element is envisaged to consist of a thin cylindrical sleeve of a compact of uranium carbide and graphite carrying a central plug and an outer sleeve, both of graphite. The latter serves to provide mechanical rigidity, to restrain the fuel in the event of physical break-down under irradiation and to increase the amount of neutron moderation. The coolant gas would flow over the surface of the outer graphite sleeve. Considerable improvement of neutron economy could be achieved by a continuous removal of the gaseous fission products. This might be accomplished by pumping out the gases through the inner plug, which would be porous; the outer graphite sleeve would be relatively impermeable to gases to prevent excessive leakage of coolant gas into the de-poisoning system and to prevent diffusion of gaseous fission products back into the coolant stream.

Such an advanced design of reactor presents many problems both of an engineering and a chemical nature. At the high temperatures involved, carbon dioxide cannot be used as coolant because of excessive reaction with graphite even in the absence of irradiation. Helium has been chosen as the most suitable coolant, but many interesting chemical problems arise, both in the purification of the gas to the required high degree and in the chemical determination of impurities at the low levels involved. Estimates of the maximum permissible concentration of such gases as oxygen, hydrogen and water vapour lie in the region of a few parts per million or less. Such low limits are necessary because of the rapid reaction between graphite and the impurities at the temperatures involved, and because the first item to be attacked would be the impervious graphite sleeve, which, while not being destroyed by only slight attack, could be rendered permeable to gases.

Although helium is thermodynamically stable in the presence of graphite, it has been suggested that in the presence of intense radiation levels He^+ ions may be produced. Such ions, being isoelectronic with the hydrogen atom, might react with the graphite to give volatile species analogous to the hydrocarbons. If the half-life for the decomposition of such species were of the order of only a fraction of a second, this would be long enough to give rise to a mass transfer of graphite out of the core. As yet, however, this interesting speculation has not been subjected to experimental investigation.

The production of an impervious graphite sleeve presents many problems. Graphite prepared in the normal way is porous, as is indicated by a comparison of the bulk density (1.65 to $1.7 \, g/cm^3$) with the theoretical X-ray density ($2.2 \, g/cm^3$). One method of decreasing the porosity is to impregnate the graphite with pitch, firing to convert the pitch to graphite and repeating the process as necessary. Alternatively the pyrolytic cracking of gaseous hydrocarbons on graphite produces a thin impervious layer upon the surface. The stability of such surface layers under intense irradiation, however, remains subject to doubt.

Little is known of the diffusion of fission products in graphite at temperatures of 1000°C and upwards. Elements which form stable carbides, such as zirconium, diffuse only slowly, as may be expected. Gases, on the other hand, diffuse rapidly through the pores of the graphite.

The diffusion coefficient of each fission product will be a function of temperature and the micro-structure of the particular type of graphite. It may depend also upon minute traces of impurity in the graphite and the diffusion properties of the precursors if it is the daughter of a short-lived product.

For dissolution of the fuel element prior to reprocessing several methods are available. Oxidation in air leaves a residue of uranium oxide containing many of the non-volatile fission products. The residue could then be dissolved in nitric acid prior to the normal solvent extraction process. The oxidation would give rise to large volumes of gases highly contaminated with the volatile fission products and with other fission products carried over as dust. Wet oxidation can be achieved with nitric acid, but the process is slow and erratic. Alternatively, the graphite may be broken down to a sludge by anodic oxidation in an electrolytic cell; most of the uranium passes into solution with the fission products.

It will be apparent that gas-cooled power reactors are far from being an obsolescent concept, and that although the prototypes at Calder Hall were out-dated even before they started to produce power, many developments have been, and will continue to be, made in this class of reactor. The chemist, metallurgist and engineer, all have vital parts to play in this development.

FURTHER READING

Jay. *Calder Hall.* Methuen and Co. Ltd., 1956.
Moore. *Development of Gas-cooled Reactors for Power.* Symposium of the British Nuclear Energy Conference, November, 1956.
United Nations. Proceedings of the First International Conference on the Peaceful Uses of Atomic Energy, Geneva, 1955. Vol. 9, *Reactor Technology and Chemical Processing.*
United Nations. Proceedings of the Second International Conference on the Peaceful Uses of Atomic Energy, Geneva, 1958. Papers 73, 81, 193, 303, 1450, 1890, 2404.

CHAPTER 9

SOME MISCELLANEOUS REACTOR SYSTEMS

THE incentives to study more exotic reactor systems are three-fold. Firstly, the use of a gas or water as coolants for reactors employing solid fuel elements is somewhat limited by the inferior heat-transfer properties of gases on the one hand and by the high working pressures experienced with high-temperature water on the other hand. Secondly, the life of a solid fuel element is limited, not only by the distortion which occurs under irradiation but also by the consumption of the fissile material. Such difficulties can, in principle, be circumvented by employing a liquid fuel. By their very nature liquids are free from any possibility of physical damage, and the fissile material burnt during operation can be renewed without any processing of the fuel charge. Thirdly, for specialised applications, for example aircraft propulsion, other factors such as power-to-weight ratio must be considered besides overall economy, and a novel reactor system may prove to be more suitable in these respects.

LIQUID-COOLED REACTORS

Two types of coolant which are preferable to water in many respects are liquid metals, especially sodium, and organic liquids. The principle problems which arise in their application are respectively the corrosion of structural materials and chemical decomposition.

ORGANIC LIQUID-MODERATED REACTORS

A liquid hydrocarbon, containing both of the light elements carbon and hydrogen, makes a good combined moderator and coolant for a nuclear reactor. By a suitable choice of compound, high pressures at the operating temperature can be avoided. Hydrocarbons decompose under irradiation to a greater or less extent, and the most stable compounds in the required melting and boiling point ranges have been found to be the polyphenyls. These compounds have been used previously in industry as high-temperature heat-transfer fluids, and common structural materials, such as steels and aluminium, are resistant to them.

A 16 MW (heat) prototype organic liquid-moderated and cooled reactor has been constructed at Idaho, U.S.A., and it became critical in September 1957. The essential features of such a reactor are depicted in Fig. 19. The fuel elements consist of plates of fully

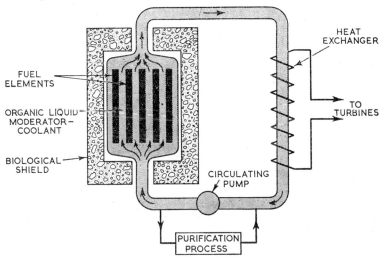

FIG. 19.   THE ORGANIC LIQUID-MODERATED REACTOR (OLMR)

enriched $U^{235}O_2$, 0·02 in. thick and clad in stainless steel 0·005 in. thick. The total critical mass of U^{235} in the core is 20 kg. The organic liquid-coolant moderator enters the core at 260°C and leaves at 370°C, and the maximum temperatures of the fuel and its sheath are 440°C and 425°C respectively.

Some relevant properties of suitable polyphenyls are summarised in Table 21.

TABLE 21.—PHYSICAL PROPERTIES OF POLYPHENYLS

	Diphenyl	o-terphenyl	m-terphenyl	p-terphenyl
Melting point (°C)	70	57	87	213
Boiling point (°C)	256	332	365	432
Vapour pressure at 325°C (at.)	3·7	0·8	0·4	0·3

Thermal Stability

The polyphenyls decompose at elevated temperatures with the

evolution of hydrogen and small amounts of methane and other simple hydrocarbons. The residues polymerise to a less volatile product, known loosely as polymer. This has not been fully analysed, but has been shown to contain higher polynuclear aromatic hydrocarbons. Fortunately, in the temperature region of interest (about 350°C) the rate of thermal decomposition is not prohibitively high. Furthermore the polymers are appreciably soluble in the original polyphenyl, and in dissolving they bring about only small changes in the physical properties of the liquid even at concentrations as high as 30 per cent by weight. The formation of polymer therefore does not seriously impair the efficiency of heat removal from the reactor, either by deposition upon important heat-transfer surfaces or by changing the properties of the coolant.

Radiation Stability

Reactor radiations (fast neutrons and gamma rays) also lead to a decomposition of the polyphenyls. Again gases, largely hydrogen, are evolved and polymer is formed. The decomposition has been studied quantitatively both by irradiation in a Van de Graaff electron beam and by small-scale experiments in nuclear reactors. The amount and composition of the gases evolved were measured by standard techniques and the amount of polymer was found by direct weighing after removal by distillation of the original polyphenyl. The results presented in Table 22 are typical.

TABLE 22.—DECOMPOSITION OF POLYPHENYLS UNDER IRRADIATION

(Total fast neutron dose $\sim 10^{17}$n. cm^{-2})

Polyphenyls	Evolved Gas (cm^3/g)	Polymer Produced (wt %)
o-terphenyl	2·0	38
m-terphenyl	1·35	33
p-terphenyl	0·38	25
4% p-, 96% m-terphenyl	0·7	34

These figures illustrate the increasing radiation stability in the series ortho, meta, para, and the stabilising effect of a small proportion of a more stable polyphenyl in the presence of one less stable. This effect, together with the fact that the melting point is reduced in binary and ternary mixtures of polyphenyls, suggests that a mixture may prove to be a more suitable coolant than a pure compound.

Such mixtures are cheaper than the pure compounds and are available under trade names such as *Santowax*.

The rate of decomposition of the coolant within a particular reactor will depend upon the detailed design of the core, as this determines the rate of energy deposition in the coolant. Typically, for each kilowatt hour of heat extracted from the core, one gram of polymer and six cubic centimetres of gas would be produced. The cost of removal of this amount of polymer and replacement by fresh polyphenyl is expected to make a relatively small contribution to the cost of the power produced. However, the feasibility of this type of reactor for large-scale power generation still remains to be proved.

Coolant Purification

The gases resulting from decomposition can be readily removed in a suitable separator and, by distillation of the coolant in a suitable by-pass circuit, the concentration of polymer in the system can be kept down to an acceptable level. Because of the very low capture cross-sections of carbon and hydrogen the induced radioactivity in the organic liquid is negligibly small. Furthermore, in the absence of corrosion, no corrosion products are present in the liquid to be activated in the reactor core and carried round the circuit. Shielding of the coolant circuit and purification plant is therefore not necessary, and the waste-product gases and polymer are inactive and require no elaborate disposal methods.

SODIUM-COOLED REACTORS

The low density alkali metals, having high specific heats and thermal conductivities, make excellent heat-transfer media. Because of their high electrical conductivity, moreover, they can be pumped electromagnetically. Here a high current is passed through the liquid as it flows between the poles of a powerful electromagnet. The relative simplicity of such pumps, which have been specially developed for the purpose, make them a much more reliable alternative to the conventional centrifugal pump.

A prototype 6·5 MW(E) sodium-cooled, graphite-moderated experimental reactor (the SRE) has been constructed in California, and became critical in April 1957. Such a reactor is depicted in Fig. 20. The fuel elements are $\frac{3}{4}$ in. diameter rods of uranium metal, enriched to 2·8 per cent U^{235}, and canned in 0·1 in. thick stainless steel tubes. A total of about 70 kg of U^{235} (or 2500 kg of the enriched uranium) is required to achieve criticality. The graphite moderator is clad with zirconium, 0·035 in. thick, on all surfaces in contact with the liquid sodium. The coolant enters the fuel channels at 260°C and leaves at 510°C.

A fast reactor must of necessity be cooled by an efficient heat-transfer medium, such as sodium, because of the small size of the core. Liquid sodium is therefore to be used as the coolant in the Experimental Fast Reactor at Dounreay; here the chemical and

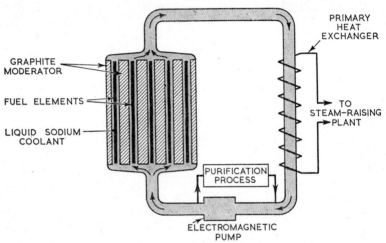

(BIOLOGICAL SHIELD SURROUNDS THE WHOLE CIRCUIT)

FIG. 20. THE SODIUM-COOLED GRAPHITE MODERATED REACTOR

metallurgical problems are similar to those arising in the graphite moderated reactor with the exception that the problem of compatibility of sodium and graphite does not arise.

Radiation Stability

Liquid metals, being fluid, are not subject to radiation damage. Sodium, however, absorbs neutrons to give the isotope Na^{24}, which decays by beta and gamma emission with a half-life of 15 hours. Consequently, it is necessary to shield not only the reactor core but also all the circuits in which the coolant flows.

Compatibility of Sodium and Graphite

Sodium does not form a carbide when in contact with graphite at temperatures below 900°C, but metals which are subject to carburation, such as stainless steels, take up carbon when in contact with both sodium and graphite at temperatures above 600°C. It is not certain whether the carbon is transferred as a dilute solution or as a suspension. Moreover, graphite is wetted by liquid sodium, and even at temperatures as low as 450°C all the available pores in the graphite are filled with liquid.

Recently evidence has been adduced for the existence of a lamellar compound of sodium and graphite. Such compounds are well known for the heavier alkali metals. On heating together powdered graphite and sodium for a few hours at temperatures between 200°C and 500°C a residue having a violet tinge is obtained. The X-ray diffraction pattern differs significantly from that of graphite. By measuring the carbon to sodium ratio in samples which showed no diffraction lines corresponding to an excess of either starting material, the formula of the compound is deduced to lie between $C_{60}Na$ and $C_{68}Na$.

This impregnation of the graphite by sodium has two consequences of interest to the reactor designer; it increases the neutron capture in the moderator by an order of magnitude and brings about a significant increase (approximately 1 per cent) in the dimensions of the graphite. Some barrier, for example a suitable metal, is therefore required to separate the sodium and graphite.

Compatibility of Sodium and Metals

Zirconium. The necessity of providing a barrier between sodium and graphite calls for a metal which is compatible with both of these materials at the operating temperature of the reactor, and which has a low neutron absorption. The latter requirement makes zirconium an obvious choice. No evidence has been found to suggest that sodium and zirconium are mutually soluble, and zirconium carbide is not produced at temperatures below 600°C.

Zirconium, however, is known to be an extremely efficient getter for oxygen. From the free energy data it is predicted that it will react with any oxygen present in the sodium, as Na_2O, to form zirconium dioxide. This has been observed experimentally as a strongly adherent film on test specimens immersed in liquid sodium containing added sodium oxide. At the reactor operating temperatures the zirconium oxide, which is known to be soluble in zirconium metal to an extent of 7 per cent, can be expected to diffuse into the metal, so destroying its mechanical strength. It is likely that during the operation of a reactor air will leak into the coolant system to an unpredictable, but presumably small, extent. Some means of keeping the oxide content of the sodium as low as possible will therefore be required if attack of the zirconium is to be avoided. Methods of achieving this are discussed below.

Stainless steel. Zirconium, while providing a method of protecting the graphite, has poor mechanical properties at elevated temperatures. Some other stronger metal, such as stainless steel, is needed for the core vessel and pipework where stresses are high. The corrosion rate of stainless steel in liquid sodium is very low. The presence of sodium oxide again results in the formation of an oxide film on the metal, but as this is insoluble in the steel no serious consequences result from its formation.

Purification of the Sodium

In view of the deleterious effects on zirconium of traces of sodium oxide it is necessary to maintain the concentration at acceptably low limits. Several methods of achieving this are available. The oxide is only slightly soluble in sodium at low temperatures, for example the solubility at 150°C is 0·004 wt. per cent, compared with 0·08 wt. per cent at 350°C and 0·2 wt. per cent at 450°C. Thus by cooling and filtering the sodium at about 150°C in a by-pass circuit the oxide content can be maintained at a reasonable level.

Alternatively, the introduction into the system of a metal which is more reactive towards oxygen than is zirconium results in the removal of sodium oxide. Calcium dissolves in liquid sodium, and readily forms an insoluble oxide which can be removed by filtration. Instead of a soluble metal, a more reactive insoluble metal, such as a 50 per cent zirconium-titanium alloy, may be used, and this is particularly effective if it is held at a temperature above that of the remainder of the circuit. This alloy removes not only oxygen but also nitrogen and hydrogen; the latter arises from reaction of the sodium with any water vapour present. Both of these gases have deleterious effects on zirconium at high temperatures. The use of this alloy also avoids the formation of a dispersion of reaction-product oxides in the sodium, with subsequent filtration problems.

LIQUID-FUELLED REACTORS

Several advantages ensue from the use of a fluid fuel in which fissile material is suspended, or preferably dissolved, and employing this same fluid as the heat-transfer medium. Heat-transfer surfaces within the reactor core are avoided, physical damage to the fuel cannot occur and it may be possible to refuel the reactor continuously and remove fission-product poisons. These advantages are offset by practical difficulties which arise, such as the possibility of chemical decomposition of the liquid and of enhanced corrosion effects in the presence of the intense radiation flux. The intense fission-product activity which is present in the whole cooling system implies that the latter requires to be heavily shielded and absolutely leak-tight.

THE HOMOGENEOUS ORGANIC MODERATED REACTOR

A solution of an organo-uranium compound in a polyphenyl, surrounded by a blanket of a solution of an organo-thorium compound would make an attractive reactor system, with many of the potentialities of the Homogeneous Aqueous Reactor and without the problems of corrosion and high pressures. It is unlikely, however, that such compounds would be sufficiently stable at the required operating temperatures of 300 to 350°C. For example, the dibenzoyl

$$\left[Th \begin{array}{c} O - C \diagup^{C_6H_5} \\[2ex] \diagdown CH \\[2ex] O = C \diagdown_{C_6H_5} \end{array} \right]_4$$

methane complexes of thorium and uranium are soluble in terphenyl, but solutions of uranyl complex undergo rapid thermal decomposition at 340°C; solutions of thorium complex precipitate thorium benzoate at 250°C. Moreover, in a homogeneous reactor all the fission fragment energy is deposited within the moderator rather than in the fuel elements and for reactors of comparable power the total energy deposited in the polyphenyl in a homogeneous reactor would be many orders of magnitude greater than in a heterogeneous reactor. The correspondingly higher rate of polymerisation therefore makes the concept of a homogeneous organic moderated reactor entirely unpracticable.

THE LIQUID-METAL FUELLED REACTOR (L.M.F.R.)

Of the many liquid metals and alloys having a low neutron cross-section and in which uranium is soluble, bismuth, with a melting point of 271°C, dissolves the greatest amount. The solubility ranges from 0·9 atom per cent at 500°C to 7·1 atom per cent at 800°C. This amount of uranium is not sufficient to produce a self-sustaining chain reaction using natural uranium with graphite as moderator. It is necessary, therefore, to fuel such a reactor with pure U^{235} or U^{233}, and in this event breeding in a blanket of thorium surrounding the reactor core is essential if power is to be produced

at an economic price.   An experimental liquid bismuth reactor
is under consideration in the United States.   A schematic diagram
of a proposed power reactor, with the associated chemical processing plants, is shown in Fig. 21.

In the proposed power reactor, fuel solution will flow through the
core—a graphite block pierced with fuel channels—at a rate of
about 1500 tons per minute.   The inlet and outlet fuel temperatures will be about 400°C and 540°C respectively.   The fuel solution
will be continuously processed in a by-pass to the main reactor
circuits, and provision will be made for the removal of the volatile
fission products and the polonium 210 produced by neutron capture
in the bismuth.

Compatibility with the Graphite Moderator

Liquid bismuth does not react with graphite over prolonged
periods, nor does it wet the graphite and percolate into the pores.
Uranium on the other hand is known to form stable carbides UC,
U_2C_3 and UC_2, and when dissolved in bismuth at a concentration
exceeding 4 per cent it has been shown to react rapidly with graphite
at 1000°C.   Visible deposits are obtained at the graphite-metal
interface and have been identified as UC by X-ray diffraction
techniques.   Such a reaction would have the serious effect of
removing fissile material from the fuel solution.   It has been found,
however, that the addition to the bismuth of small amounts of
zirconium, which forms a more stable carbide than does uranium,
brings about the preferential formation of zirconium carbide on the
graphite surface.   The strongly adherent layer so produced effectively prevents subsequent reaction between uranium and the
graphite.

Compatibility with Metals

Although liquid bismuth containing uranium and zirconium is
compatible with graphite, some metal container is required for the
core vessel and pipework.   Exhaustive tests have shown that chrome
steels are sufficiently corrosion-resistant, the penetration of the
metal being about 0·001 in. after 1000 h at 550°C.   The choice
of container material is somewhat limited, because it has been
shown that when some metals dissolve in liquid bismuth, to an
extent which corresponds to only a small degree of corrosion,
precipitation of the uranium as an intermetallic compound occurs.
For example, nickel precipitates the uranium as UNi_2 and UNi_5.

Although chrome steels are adequately corrosion-resistant under
isothermal conditions, when thermal gradients exist in a flowing
system mass transfer of iron can take place.   Metal is removed

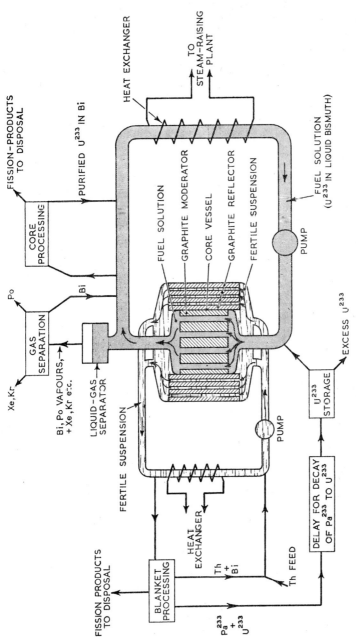

FIG. 21. A PROPOSED LIQUID-METAL FUELLED REACTOR

Fuel—solution of U²³³ in liquid bismuth
Fertile material—suspension of Th₃Bi₅ in liquid bismuth.

from the hotter parts of the system and is deposited, again as metal, in the cooler parts. This is attributed to the variation with temperature of the solubility of the metal in bismuth.

Laboratory investigations of mass transfer have been carried out using small thermosyphon loops, in which a tube of the material under test is fabricated into a parallelogram and held in a vertical plane. One upright limb is heated and the other cooled, so that flow takes place due to the difference in the density of the liquid in the two limbs. For given rates of heating and cooling, the temperature difference between the limbs is a measure of the liquid flow-rate. As the cold limb begins to block with mass-transfer deposit, flow is restricted and the temperature difference increases. Typical operating temperatures are 550°C in the hot limb and 400°C to 520°C in the cold limb, giving a flow-rate in a loop of 1 in. bore of up to a few ft/sec. Blockage of the loops by a crystalline deposit has been observed after periods of operation varying from hundreds to tens of thousands of hours, depending upon the type of steel employed.

The addition of zirconium to the bismuth inhibits the process of mass transfer to a striking degree, concentrations as low as 5 p.p.m. being effective. This inhibition is attributed to the deposition of zirconium on the surface of the steel. It has been shown by radioactive tracer methods that such a deposit not only reduces the rate of dissolution of the iron, but also interferes with the deposition of the crystallites of iron in the cooler parts of the system. The crystallites are therefore carried round into the hot zone, where they redissolve, in preference to further dissolution of the vessel wall.

Chemical Processing of the Core

Removal of the fission products from the fuel of a liquid bismuth reactor can be achieved, at least in part, by a high-temperature liquid-extraction process. When liquid bismuth, now containing fission products as well as uranium and zirconium, is contacted with a fused mixture of alkali chlorides, such as $LiCl + KCl$, displacement reactions take place. The reaction of lanthanum is a typical example:

$$La + 3LiCl \rightarrow LaCl_3 + 3Li$$

The fission-product chlorides produced in this way dissolve in the salt phase, which is subsequently separated from the purified bismuth. The free energies of formation of many of the fission-product chlorides are such that, even when the extracted metal is at a low concentration in the bismuth phase, the high concentration of lithium chloride drives the reaction nearly to completion.

Fortunately the stability of uranium tetrachloride is low so that uranium is not extracted. The method can be further improved by dissolving in the fused salt mixture both magnesium metal and magnesium chloride. By varying the relative proportions of these two components the oxidising power of the melt can be adjusted as required, and a wider range of fission products can therefore be extracted.

The Fertile Blanket

In order to achieve a high degree of conversion, the reactor core must be surrounded by Th^{232} at a high concentration. Thorium has only a slight solubility in bismuth at the temperatures of operation and the blanket was originally envisaged as a slurry in bismuth of the compound Th_3Bi_5. Processing would involve dilution of the slurry with bismuth to give a homogeneous solution of thorium, containing also Pa^{233}, U^{233} and fission products, followed by a fused-salt extraction process similar to the one planned for the core.

Experiments with thermosyphon loops have revealed two physical phenomena which at present are a serious handicap in the development of a slurry blanket. A redistribution of the sizes of the slurry particles has been observed, there being a strong tendency for a few large particles to grow at the expense of smaller ones. Moreover, deposition of particles on the walls of the cooler parts of the system occurs, and such deposits are strongly adherent. Both these effects would lead to considerable difficulties in pumping the slurry.

As an alternative to a metallic slurry blanket a slurry of thoria in heavy water has also been suggested; the problems which would arise here are very similar to those discussed in Chapter 7 for the blanket of the Homogeneous Aqueous Reactor.

FUSED SALT SYSTEMS

Fused salts are becoming more widely used in industrial processes both as media for heat-treatment baths and for heat transfer at elevated temperatures. The three types of salt which have been considered either as reactor coolants or as carriers for fissile material in a fluid-fuel reactor are:

(a) Nitrate-nitrite mixtures, e.g., the $NaNO_3$-$NaNO_2$ eutectic, m.p. 220°C.

(b) Hydroxides, e.g., NaOH, m.p. 320°C.

(c) Fluorides, e.g., $NaF : ZrF_4$ (50 : 50 mole %), m.p. 550°C.

The choice of compounds is limited to these three groups by considerations of thermal stability, neutron absorption and melting point. Desirable features in any liquid to be used as a carrier for fissile material are stability under irradiation, a high solubility of uranium and compatibility with structural materials.

Radiation Stability

Hydroxides and fluorides, being largely ionic in character, might be expected to be stable under irradiation, and this is borne out by experiment; only slight amounts of gas are evolved under irradiation intensities of several hundred kW/l of melt. Nitrates, on the other hand, having covalent bonds, are unstable towards irradiation. Under gamma and fast neutron irradiation in the BEPO reactor, samples of potassium nitrate evolved significant amounts of oxygen, with a corresponding reduction of nitrate to nitrite. An equilibrium overpressure of oxygen was obtained, showing that the recombination reaction between oxygen and nitrite is of importance under the experimental conditions.

Solubility of Fissile Material

The solubility of a uranium (or plutonium) compound can be determined in the usual manner by equilibrating the melt with the compound and sampling the liquid phase for analysis. The sampling may be carried out either by filtering the melt, under inert gas pressure, through a sinter of a suitable metal or graphite, or by centrifuging the mixture at the required temperature. The latter technique, developed at Harwell, is of particular value in measuring small solubilities.

Uranium compounds have been found to be almost insoluble in both fused caustic soda and in potassium nitrate-nitrite mixtures. For example, the solubility of UO_3 in NaOH is 0·01 wt. per cent at 400°C rising to 0·03 wt. per cent at 650°C. The insoluble phase consists of a sodium uranate of which two modifications have been identified: a red compound, Na_4UO_5, produced at low concentrations of UO_3, and a yellow compound, with a composition in the range $Na_2U_2O_7$ to $Na_2U_4O_{13}$, produced at higher concentrations of UO_3. All uranium compounds, including U(VI) and the metal, are converted to the insoluble U(VI) uranates in contact with molten caustic soda under a non-reducing atmosphere. This interesting oxidising property of caustic soda is further illustrated in its reactions with metals which are discussed below.

Uranium tetrafluoride (m.p. 1030°C) forms liquid binary and ternary mixtures with many other fluorides, and some of these have acceptably low melting points. For example, with NaF (m.p. 1000°C) a binary eutectic is formed of composition 76 mole per cent NaF, 24 mole per cent UF_4 with a melting point of 610°C. With

NaF and PbF_2 (m.p. 840°C) a ternary eutectic, of melting point 470°C and composition 31 mole per cent NaF, 63 mole per cent PbF_2, 6 mole per cent UF_4 is formed. In the equimolar binary eutectic of NaF and ZrF_4 (sublimes 1200°C) the solubility of UF_4 increases from zero at the melting point of the eutectic (550°C) to several mole per cent at 600°C.

Compatibility with Structural Materials

Mixtures of nitrates and nitrites are used in industry as heat-transfer media, and the working temperature is limited to below 600°C by the thermal decomposition of the nitrate. Mild steel is a suitable container material, but where heavy duty is required, for example in heating element sheaths, nickel or high-nickel alloys are preferable.

Molten caustic soda reacts with many metals with the evolution of hydrogen:

$$6NaOH + 2Fe \xrightarrow{600°C} 3Na_2O + Fe_2O_3 + 3H_2$$

Surprisingly enough, aluminium is inert below its melting point (660°C), presumably due to the formation on the metal of an insoluble adherent film of sodium aluminate.

With certain metals further reactions take place, particularly at temperatures above 600°C. Nickel in particular has been extensively studied, and this is the most suitable commercially available material for containing caustic soda up to 600°C. Initially, nickel oxide is produced:

$$Ni + 2NaOH \longrightarrow NiO + Na_2O + H_2$$

and under vacuum, when the hydrogen is removed, only this reaction takes place. In a closed system, however, the extent of corrosion is reduced, and an equilibrium overpressure of hydrogen is obtained. Further reactions are postulated to account for the setting up of this equilibrium:

$$NiO + H_2 \rightarrow Ni + H_2O$$
$$Na_2O + \tfrac{1}{2}H_2 \rightarrow NaOH + Na$$
$$Na + H_2O \rightarrow NaOH + \tfrac{1}{2}H_2$$

Under these conditions corroded nickel is deposited from the melt as crystals of metallic nickel, rather than as the oxide.

While these corrosion reactions of nickel are not serious in an isothermal system below about 600°C, in a circulating, non-isothermal system undesirable mass-transfer can occur at temperatures as low as 500°C. The nature of the deposit in the cooler region depends upon the atmosphere above the liquid. In an oxidising atmosphere, or under vacuum, the deposit consists of nickel oxide, while under hydrogen the metal is deposited. A reducing atmosphere minimises the extent of mass transfer, but even under hydrogen the maximum temperature of operation is limited to about 600°C.

Fluorides vary considerably in their corrosive properties, and some measure of understanding can be gained from a consideration of the free energies of formation of the fluorides involved in the corrosion reactions. Thus melts containing unstable fluorides, such as PbF_2, are more corrosive than those containing the more stable fluorides, such as ZrF_4. Similarly, metals whose fluorides are unstable make suitable container materials. At 750°C the free energy changes (per mole of nickel fluoride produced) for the two reactions

$$PbF_2 + Ni \rightarrow NiF_2 + Pb$$
$$\text{and } \tfrac{1}{2}ZrF_4 + Ni \rightarrow NiF_2 + \tfrac{1}{2}Zr$$

are $- 2$ kcal and $+ 64$ kcal respectively. Lead is deposited from the molten $NaF\text{-}PbF_2$ mixture in a nickel crucible, whereas the molten $NaF\text{-}ZrF_4$ mixture becomes only slightly contaminated with NiF_2 (less than a few p.p.m.).

The addition of uranium tetrafluoride has an interesting effect on an otherwise non-corrosive melt, for it is quite readily reduced to the trifluoride, with corresponding oxidation of the containing metal. For example, while chromium is barely attacked by pure $NaF\text{-}ZrF_4$ melts, the addition of UF_4 results in corrosion taking place:

$$Cr + 2UF_4 \rightarrow CrF_2 + 2UF_3 \text{ (in } NaF\text{-}ZrF_4)$$

The melt can become contaminated with several thousand p.p.m. of CrF_2. In the almost complete absence of any information on the magnitude of activity coefficients of species dissolved in fluoride melts, accurate predictions of the extent of corrosion in any given system cannot be made. Approximate estimates indicate that graphite and nickel-molybdenum alloys should be resistant to attack by the $NaF\text{-}ZrF_4\text{-}UF_4$ melt, and these indications are borne out by experiment.

The activity coefficients of the components in a series of $NaF\text{-}ZrF_4$ mixtures have been obtained by measuring the partial vapour pressure of the components using the transpiration method. Compound formation occurs in both liquid and vapour phases. At 900°C, in an equimolar mixture, the activity coefficient of ZrF_4 is 0·1, and the measured vapour pressure over the composition range 50 to 100 mole per cent ZrF_4 can be interpreted in terms of an almost ideal mixture of ZrF_4 and the compound $NaZrF_5$.

In principle it should be possible to determine activity coefficients and free-energy changes for corrosion reactions by the measurement of the e.m.f. of galvanic cells with the melt as electrolyte. In practice this has proved to be very difficult, largely because of the absence of any electrical insulator which is completely resistant to the melts, and because of the unknown liquid-junction potentials which arise.

Of these fused-salt systems, therefore, the fluorides appear to be the most promising as liquid fuels for a high-temperature reactor. The low solubility of fissile material in nitrates and hydroxides would involve the use of a slurry, with the attendant problems of maintaining a uniform concentration. Moreover the corrosion by hydroxides at temperatures above 600°C and the radiation decomposition of nitrates further limit their usefulness.

Much development work on fluorides is required before they can be seriously considered as a reactor fuel, especially on the corrosion aspects and on the influence of radiation on corrosion rates. In 1954 an experimental circulating-fuel fused-fluoride reactor was

TABLE 23.—THE PRESENT STATE OF REACTOR DEVELOPMENT

Reactors Producing Useful Power			
Type	Country	Power (MW)	Date
Gas cooled, graphite moderated	U.K.	38E	1956
Improved gas cooled	U.K.	150–250E	1960
Improved gas cooled	France	60E	1959
Graphite moderated, water cooled	U.S.S.R.	100E	1958
Light water moderated/cooled (submarine)	U.S.A.	—	1954
Light water moderated/cooled	U.S.A.	60E	1957

Prototype and Experimental Power Reactors			
Type	Country	Power (MW)	Date
Graphite moderated, water cooled	U.S.S.R.	5E	1954
Boiling water reactor	U.S.A.	5E	1957
Heavy water moderated/cooled	Canada	20E	1961
Homogeneous aqueous reactor—two zone	U.S.A.	5H	1957
Graphite moderated, sodium cooled	U.S.A.	6·5E	1957
Sodium cooled fast reactor {	U.S.A.	0·2E	1951
	U.K.	60H	1959
Organic liquid moderated and cooled	U.S.A.	16H	1957
Fused salt	U.S.A.	2H	1954

Power output: H = Heat, E = Electrical.

operated for a matter of days at the Oak Ridge National Laboratory, developing a maximum power of 2 MW at a temperature of 800°C. The experiment was designed to demonstrate the feasibility and to study the operational behaviour of such a system, and was not intended to run for a long period. While the pumping of molten salts at these temperatures is no mean technological achievement, even in the absence of an associated nuclear reactor, it is still a long design extrapolation to a reliable operational reactor required to last for many years.

The Present State of Reactor Development

In conclusion it is of interest to summarise the types of reactor which have been operated or are under construction. In Table 23 the date at which each type of reactor became, or is expected to become critical is listed, along with the power output. The Table is intended to serve as a guide to the relative stage of development of the various reactor systems which have been discussed in the last three chapters, and is in no way an exhaustive list of all possible types.

FURTHER READING

Charpie, Hughes, Littler and Trocheris. Progress in Nuclear Energy, Series II. *Reactors*, Vol. 1.   Pergamon Press 1956.
Davidson, Loeb and Young. "Nuclear Reactors for Power Generation." *Annual Review of Nuclear Science*, Vol. 6, p. 317, Annual Reviews, Inc., 1956.
"The Aircraft Reactor Experiment." *Nuclear Science and Engineering*, Vol. 2, No. 6, pp. 795 to 853, Academic Press, Inc., New York, November 1957.

CHAPTER 10

THE DISPOSAL OF RADIOACTIVE WASTES

JUST as the ash, soot and noxious gases are an embarrassing waste-product of conventional sources of heat, so also are the intensely radioactive fission products an embarrassment in the operation of nuclear reactors. For over a century the combustion products of coal have been discharged into the atmosphere in ever-increasing amounts, and only in recent years has any effective legislation been introduced to control the discharge. By contrast the danger to human life resulting from the indiscriminate use of radioactive materials is well known, and so from the very beginnings of the industry extremely careful measures have been taken to ensure that radioactive wastes are dealt with in such a way that the health of the community is not endangered.

Whatever method of dealing with the radioactive waste is used, two basic requirements must be fulfilled:

1. The method must be safe, so that under no circumstances can any person either be exposed to dangerous radiation arising from the fission products or ingest a harmful amount of them.

2. The method of disposing of fission products must be economical, as the cost of processing and disposal must be charged against the electricity sent out by the nuclear power station.

Although the greatest amounts of radioactive wastes arise at the plants for processing spent fuel elements, other establishments produce wastes of lower activity which must be suitably processed before discharge into the sea or a river.

PRESENT METHODS OF WASTE DISPOSAL

The various establishments of the U.K.A.E.A. each produce radioactive wastes of some kind, which are dealt with in a manner which depends on the nature of the waste.

GASEOUS WASTES

There are two main sources of gaseous radioactive waste. Firstly, in air-cooled reactors (e.g. BEPO or Windscale) activity arises from neutron capture in the coolant as it passes through the reactor. Fortunately the only isotope produced in any significant amount is A^{41} (formed by the reaction A^{40} $(n, \gamma) \rightarrow A^{41}$); other constituents of air either have a very low cross-section or the active products formed decay very rapidly. The amount of activity produced is so small that no hazard is produced when the outlet gases from the reactors are discharged into the atmosphere through a tall stack. Great care is taken to filter the coolant, however, to ensure that any activated dust carried by the gas-stream is not discharged into the atmosphere.

Secondly, when irradiated fuel elements are dissolved prior to processing the gaseous fission products, Xe and Kr, are released. Here again the activity at present involved is sufficiently low to permit the controlled discharge of these gases into the atmosphere.

SOLID WASTES

Wastes of high activity, such as equipment highly contaminated with plutonium, can be cast into concrete blocks in a mild steel shell and dumped at sea in depths exceeding 1500 fathoms. The rate of distruption of these units will be extremely slow, and so it is ensured that the activity will be released slowly, if at all, and be dispersed without building up local high concentrations. When practicable, combustible waste is reduced in bulk by incineration prior to disposal. Dumping at sea, though costly, is convenient for the safe disposal of the comparatively small amounts of active solid waste which arise.

LIQUID WASTES OF LOW ACTIVITY

These arise at most of the U.K.A.E.A. establishments in very large quantities and are treated in much the same way as any normal trade waste. The object of the method of treatment is to transfer the extremely diluted radioactive material from the large volume of effluent into a small bulk of solid material. When the activity in the liquid has been reduced to a sufficiently low value the effluent can then be discharged into a river or the sea by agreement with the appropriate Government or Local Authority. Only minute concentrations of many radioactive materials can be permitted in the purified effluent. For example, an impurity of a few p.p.m. is usually acceptable in normal industrial effluents, but at Harwell, where the effluent is discharged into the River Thames, the concentration of certain radioactive isotopes must be reduced to the region of 10^{-12} p.p.m. or even less.

At these low concentrations it is not possible to employ direct precipitation for removal of the impurities. The activity is therefore removed by carrying out a precipitation of calcium phosphate, in the presence of ferric phosphate which improves the settling properties of the precipitate. Much of the activity is co-precipitated with the mixed calcium-ferric phosphate, but the supernatant contains some Cs^+ (which, being monovalent, is not strongly absorbed on the precipitate) ruthenium (which can exist in solution as anionic or uncharged species) and radioactive colloidal particles. Precipitation of FeS by the addition of $FeSO_4$ and Na_2S removes some of the caesium and much of the colloidal material. The effluent is finally purified by passage through an expendable ion-exchange column of the mineral vermiculite. The sludges and exhausted vermiculite are disposed of as solid waste.

At Springfields and Capenhurst the radioactivity in the effluent arises largely from uranium and its daughters; these are removed by the precipitation of ammonium diuranate. The specific activity of natural uranium is so low that the maximum amount which may be discharged is fixed by economic limits rather than on the grounds of public health.

At Windscale the low activity effluent, after a treatment similar to that used at Harwell, is discharged by pipeline into the sea. One of the conditions which determined the siting of the Windscale processing plant was the suitability of the tidal currents for the off-shore disposal of effluent. Before construction of the plant was started extensive tests were carried out, using fluorescein dye, to determine the optimum outlet points for the waste pipelines, and to ensure that under no circumstances would unfavourable conditions of tide and wind give rise to build-up of activity on the sea-shore. The situation is complicated by the possibility of concentration of certain isotopes by deposition on sand and sea mud and by incorporation in the metabolism of marine life, especially fish (via algae)

TABLE 24.—TYPICAL MAXIMUM PERMISSIBLE DISCHARGE
OF FISSION PRODUCTS OFF-SHORE FROM WINDSCALE

Isotope	Factor Limiting Daily Discharge	Max. Permissible Daily Discharge (Curies)	
		(a) Original Estimate	(b) Level Fixed by Practical Experience
Sr^{90}	Fish and edible seaweed	1	160
Ru^{106}	Fish and edible seaweed	100	65
Total β (less Ru^{106})	}Edible seaweed	50	500
Pu^{239}	Fish	0·04	2

and sea-weed, which is eaten as lava bread. Taking into account these concentration effects, the maximum levels of daily discharge of the various isotopes were laid down; after several years' operating experience it was found that these values were too conservative and could be relaxed somewhat. Typical values of the maximum permissible discharge are given in Table 24; the actual values depend upon the relative proportions of each isotope discharged.

Comparison of these figures with the daily output of fission products from a single power station (for instance, 2000 curies Sr^{90}) shows that only a very small proportion of the total output can be discharged in this manner.

LIQUID WASTES OF HIGH ACTIVITY

These arise only from the main extraction columns of the plants for processing spent fuel elements, and contain the major portion of the fission products produced during the irradiation in a reactor. The effluent contains, besides the fission products, varying amounts of other materials, including:

(a) Reagents added during the chemical process.

(b) Products arising from corrosion of the plant.

(c) Materials present in the irradiated fuel element and not removed in the process; for example, impurities present in the original fuel, alloying and cladding materials and species produced during the irradiation (e.g. Np, Am, etc.).

(d) Reactor fuel which is incompletely removed (U, Pu).

The high activity effluent from a solvent extraction plant such as Windscale has the approximate composition shown in Table 25.

TABLE 25.—APPROXIMATE COMPOSITION OF THE
MAIN EFFLUENT FROM THE WINDSCALE SOLVENT
EXTRACTION PLANT

Component	Concentration (g/litre)
Nitric acid	200
$Fe(NO_3)_3$ (corrosion product)	0·6
$Cr(NO_3)_3$ (,, ,,)	0·2
$Ni(NO_3)_2$ (,, ,,)	0·1
$Al(NO_3)_3$ (canning material)	0·1
$UO_2(NO_3)_2$	0·2
Fission product nitrates	0·2
[or fission product cations	0·1]

The exact composition depends upon the details of the extraction process. For example, the Windscale effluent has a high concentration of nitric acid, since this is used as a salting-out agent. On the other hand in the Redox process, which is used in some U.S. plants, aluminium nitrate is used for salting-out, so the effluent there contains a high concentration of this reagent.

The high concentration of materials other than fission products in the solution is noteworthy, for in some instances it is this which limits the applicability of otherwise attractive alternative methods of disposal which are being investigated.

The detailed chemical composition of the fission products themselves is complex, and, besides being dependent upon the irradiation history of the fuel element, it changes with time because of the chemical transmutations which continuously occur as a result of beta decay. Average values of the relative proportions of the individual fission products after a decay of thirty days are given in Table 26.

TABLE 26.—COMPOSITION OF FISSION PRODUCTS AFTER 30 DAYS' COOLING

Element	Percentage of Total Weight
Rare earths and yttrium	30
Inert gases, Xe + Kr	15
Zirconium	14
Caesium	11
Molybdenum	10
Ruthenium, rhodium, palladium	6
Strontium	4
Barium	3
Technetium	3
Tellurium	2
Rubidium	1
Niobium	0·5
Iodine	0·5
	100

The high activity wastes are stored as an aqueous solution of nitrates in large tanks, and because of the long half-lives of some of the fission products, storage must be for an almost indefinite period. Of necessity the tanks must therefore be virtually indestructible, and elaborate precautions are necessary to ensure the early detection of any leakage. In order to minimise the storage capacity required the effluent is evaporated prior to storage. This process introduces

M 2

several chemical problems ; the results of laboratory investigations on some of the problems which can arise have been published.

Small amounts of organic solvent are carried over in the active effluent, and evaporation of the solution, with a consequent increase in the nitric acid concentration, could lead to violent reaction between the organic material and the acid. Stripping of the effluent with steam prior to evaporation effectively removes organic contaminants.

The object of the distillation is to reduce the solution of fission products to a small bulk and to dispose of the distillate as low-activity waste. Carry-over of activity in the distillate, either as spray or by the formation of volatile fission product compounds, must therefore be avoided. In a solution of hot concentrated nitric acid radioactive ruthenium, which is usually found in the wastes as the non-volatile reduced forms, is oxidised to the volatile ruthenium tetroxide. From a solution 15 N nitric acid, ruthenium can be quantitatively distilled by evaporation to near dryness. The rate of oxidation, and hence of distillation, of ruthenium increases with increasing acid concentration and with increasing degree of evaporation. The distillation of ruthenium can be reduced to very small proportions by evaporation at a reduced pressure, and hence at a lower temperature. Even when evaporation is carried out to saturation of the residue, less than 10^{-4} per cent of the ruthenium then appears in the distillate.

Nitric acid can be removed from the wastes without introducing additional electrolytes by the addition of a suitable reducing agent, which is chosen so that the products of the reduction are all gaseous. Formic acid, formaldehyde and nitric oxide are all suitable for this purpose. As in the familiar copper-nitric acid reaction different mechanisms apply at different acid concentrations:

In strong acid (8 to 16 N) $2HNO_3 + R \rightarrow RO + 2NO_2 + H_2O$
In dilute acid (1 to 8 N) $2HNO_3 + 3R \rightarrow 3RO + 2NO + H_2O$

Here RO represents the oxidation products of the reducing agent R, and in the case of formaldehyde or formic acid $RO = CO_2 + H_2O$ and for nitric oxide $RO = NO_2$, hence nitric oxide is effective only in reducing the stronger acid. Below an acid concentration of 1 N, formaldehyde is oxidised to formic acid only. The reactions of the organic reagents are very rapid at temperatures around 100°C, and are catalysed by both uranyl and ferric ions, both of which are present in the high activity effluent.

A considerable saving of formaldehyde or formic acid could be achieved by using the nitric oxide evolved in the reduction of the weaker acid to reduce a second batch of stronger acid. Thus using formaldehyde alone as the reducing agent two moles of nitric acid are reduced per mole of formaldehyde:

$$4HNO_3 + H.CHO \rightarrow 4NO_2 + CO_2 + 3H_2O \text{ (in } > 8N \text{ acid)}$$

followed by:

$$4HNO_3 + 3H.CHO \rightarrow 4NO + 3CO_2 + 5H_2O \text{ (in } < 8N \text{ acid)}$$

Using the nitric oxide evolved in the second of these reactions to reduce the nitric acid in the next batch of strong acid,

$$4NO + 5\tfrac{1}{3} HNO_3 \rightarrow 9\tfrac{1}{3} NO_2 + 2\tfrac{2}{3} H_2O$$

approximately three moles of nitric acid are reduced for every mole of formaldehyde consumed.

PROBLEMS TO BE FACED IN AN EXPANDING NUCLEAR POWER PROGRAMME

MAGNITUDE OF THE PROBLEM

Approximately 3 or 4 grams of fission products are produced each day for every megawatt of electricity generated in a nuclear power station. Depending on the reactor type, the fission products are removed either continuously (as in a liquid fuelled reactor), or

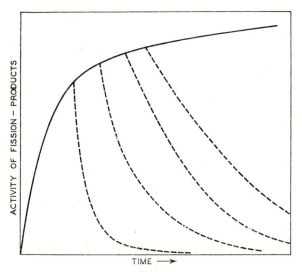

FIG. 22. GROWTH OF FISSION-PRODUCT ACTIVITY DURING OPERATION OF A TYPICAL NUCLEAR REACTOR (FULL CURVE) AND DECAY ON SHUT-DOWN (DOTTED) AFTER VARIOUS OPERATING TIMES

periodically, when solid fuel elements are reprocessed. The activity of the fission products in a reactor grows with time, as shown schematically in Fig. 22. The actual shape of the curve depends upon the type of reactor, its power level and other details. In the figure, the initial steep rise is due to short-lived fission-product isotopes, which soon reach a state of equilibrium and then decay as rapidly as they are formed; the less rapid build-up is due to the accumulation of isotopes of longer half-life. This build-up is further illustrated by the manner in which the fission-product activity dies away on shut-down of the reactor after various operating times (dotted curves). With increasing irradiation times the initial rate of fall-off of activity is less rapid, and the proportion of longer-lived activity increases.

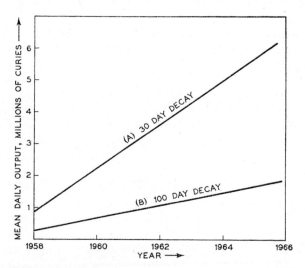

FIG. 23. ESTIMATED MEAN DAILY OUTPUT OF FISSION-PRODUCT ACTIVITY (BETA PLUS GAMMA) ASSUMING DECAY TIMES OF 30 AND 100 DAYS

A similar curve can be constructed for a series of power stations, based on the installation of 6000 MW of nuclear power by 1965, as outlined in the Government White Paper (March 1957). Assuming a uniform rate of introduction of generating capacity over the period considered, the mean daily output of fission-product activity may be calculated and is plotted in Fig. 23; curves A and B represent the daily output assuming that the fission products are allowed to decay for 30 days and 100 days respectively before disposal. It can be seen that one very effective method of reducing the quantity of activity to be disposed of is simply to store the fission products for as long a time as possible. A storage of 100 days is not unreasonable, and, in fact, is often desirable prior to chemical processing by solvent extraction, as too great an activity in the process liquors can lead to difficulties due to the radiolytic decomposition of the solvents.

The growth with time of the total amount of fission-product activity arising from the national power programme is shown in Fig. 24. Only the activity with a half-life of greater than two

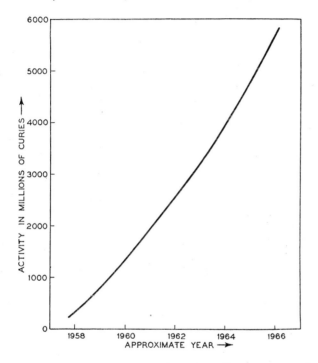

Fig. 24. Estimated Accumulation of Long-lived Fission-product Activity Arising from an Expanding Power-programme

(Only fission-products with a half-life of greater than two months are considered.)

months is considered, because the fission products are not normally separated from the fuel elements before this order of time has elapsed. The amounts of activity produced in this extensive power programme appear formidable, especially when compared with the maximum permissible body levels of many of the isotopes involved (of the order of 10^{-5} c) or with the activity required to produce the maximum permissible level of radiation. For example, an un-shielded source of one million curies of fission products would give the maximum permissible radiation level at a distance of several miles!

NATURE OF THE PROBLEM

The Atomic Energy Industry is therefore faced with the problem of disposing of a steadily increasing output of highly active fission products. As the plant at Windscale was built primarily to produce plutonium for weapons, speed rather than cost was therefore of the greatest importance and the methods of dealing with the highly active waste are not necessarily the cheapest and best. Two methods of approach to the problem are in principle possible; the fission products can either be dispersed throughout a large volume, in which the concentration is so low that no danger is involved, or be stored at high concentration in such a way that escape of the activity is made impossible, as at present.

We have seen that only a very small proportion of the anticipated output can be discharged as liquid waste off-shore, and unless the fission products are in a very concentrated form, and hence occupy only a small volume, transport costs for deep-sea dumping become expensive. Furthermore, once dumped at sea no further control over any movement of the activity is possible, so that as a means of disposing of the very large quantities of activity which will be produced in the future this method is internationally undesirable. Likewise, dispersal into the atmosphere is suitable only for relatively small amounts of activity; apprehension has already been expressed concerning the genetic effects of the activity released by weapons tests. These to date have probably distributed a total of the order of 1000 kg of fission products, compared with an anticipated annual output from United Kingdom power reactors of about 5000 kg by 1965. We are left, therefore, with storage as the most acceptable method of waste disposal, and means of converting the fission product wastes into a form suitable for storage are being actively pursued in many countries.

DEVELOPMENTS IN THE TREATMENT OF HIGHLY ACTIVE WASTES

Although new methods of processing, such as were described in Chapter 5, may be introduced at some time in the future, it is likely that the solvent extraction process will continue to be important for many years. For our present purposes, therefore, discussion will be limited to the disposal of aqueous effluent, although several of the methods described do lend themselves to disposal of solid wastes.

CALCINATION

Perhaps the most obvious method of concentrating the fission products is by evaporation to dryness and calcining the resulting mixture of nitrates to oxides. The later stages of the calcination, however, give rise to a considerable dust hazard, and the residue is a friable powder which is difficult to handle and must be stored in hermetically-sealed containers.

ION EXCHANGE

Many minerals, and especially clays, exhibit strong cation-exchange properties, and will take up fission products from an aqueous solution. Water is eliminated during subsequent firing of the clay and the ion-exchange properties are lost. Thus, on passing a solution of fission products through a column of montmorillonite, about one gram-equivalent of cations is taken up per kilogram of clay, and after firing at 1000°C less than 0·1 per cent of the activity is removed after several months of leaching with either distilled water or brine.

A difficulty is encountered in preparing a suitable column on which to carry out the exchange. The particles of clay are so fine that a settled bed is almost impervious, but by extruding a thick mastic into water a suitable column can be prepared; this tends to break down over extended periods of time, however. A more suitable method is to bond the clay with tetraethyl orthosilicate, and then decompose the binding agent by heating to give a matrix of silica. The resulting product can then be ground and sieved to any desired size. Although the heating process reduces the ion exchange capacity by about one-half, the clay still retains its property of fixation of cations on firing. The maximum concentration of fission products obtainable is about 50 to 100 g/l of fired clay. This is comparable with the concentration attainable in aqueous storage vessels.

CALCINATION FOLLOWED BY ION EXCHANGE

A serious disadvantage of the ion exchange method when applied to the effluent of a solvent-extraction plant arises because of the presence of cations other than fission products, (Al, U, etc., see Table 25). These cations are also adsorbed on the columns, and therefore reduce by as much as 90 per cent the effective capacity for fission products. The interfering elements can be removed by evaporation of the solution to dryness and calcination at a suitable temperature, when the nitrates of aluminium, uranium, etc., are converted to oxides, from which some of the fission products are leached by water. For example, on calcination at 700°C, all the activity except that arising from the alkali and alkaline earth cations is firmly held against leaching; the leachable constituents are absorbed on a small ion-exchange column and fixed in the usual way.

This process suffers from the disadvantages of direct calcination, and also would be cumbersome to perform under highly active conditions.

CONVERSION TO GLASS-LIKE MATERIALS

An alternative method of avoiding the difficulty of the overloading of ion-exchange columns by an excess of inactive cations is to mix the waste solution with some fusible material, evaporate to dryness and fire to give a glass-like residue. There are two variations of the method:

(a) *Nitrate Systems.* If the high nitric acid content of the effluent is neutralised with caustic soda, sufficient sodium nitrate is produced to provide a suitable carrier for the solid wastes. After evaporating to dryness and fusing at 370°C, the melt is cast into trays for storage. The trays are shallow to permit the escape of the fission-product heat without causing fusion with subsequent release of the volatile fission products. The trays must be stored in such a way that leaching by water is impossible, since many fission-product nitrates are soluble in water.

(b) *Silicate Systems.* Glasses are known to be highly resistant towards leaching of their constituent cations, so that incorporation of fission products into a glass should yield a compact product in

TABLE 27.—LEACHING FROM SOILS FIRED AT 1000°C
(From "Treatment of Highly Active Wastes" by C. B. Amphlett, *Atomics*, April 1957)

	1	2	3
Solution treated	$CsNO_3$ 0·05N $Fe(NO_3)_2$ 3·0N	$Cr(OH)_3$, $Fe(OH)_3$ (slurry)	Synthetic waste solution
Loading (equivalents per kg of product)	1·7	11·5	2·0
Flux	None	None	$NaNO_3$
	Friable	Friable	Glass
Per cent leached by water	0·05 (1 week)	0·001 (1 week)	0·005 (2 weeks)

which the activity is firmly retained. For economic reasons it is obviously desirable to use a locally available silicous material with which to prepare the glass, and work has been carried out at Harwell, Oak Ridge and Chalk River using in each case local soils.

Work at Harwell has shown that on slurrying the soil—green-sand—with a fission-product solution, drying and firing at 1000°C, a sintered mass is obtained. The mass is friable, but is very resistant to leaching. On firing to 1400°C the sinter fuses to give a dense glass, which is also resistant to leaching. The melting point of the soil can be reduced by the addition of fluxes such as sodium carbonate, nitrate or borate without seriously impairing the leaching properties (Table 27).

Interesting features of the process are:

(a) The high loadings which are achievable; a fission-product density of at least 400 g per litre of glass, compared with about 100 g per litre conveniently attainable in stored aqueous solutions.

(b) The versatility of the method; solid wastes can be dealt with in the same manner as liquid wastes.

(c) The method is unaffected by the presence of a very large excess of inactive cations.

The firing could be carried out without any external heating. Calculations have shown that, if the dimensions of the bed are sufficiently large so as to avoid excessive heat loss, the heat developed by the radioactive decay of the fission products is sufficient to bring about the fusion of the soil.

STORAGE OF THE FIXED FISSION PRODUCTS

The bed of fission products, fixed by one of the methods described above, must be stored, preferably near to the separation plant for economic reasons, in such a way that any slight leaching does not have harmful consequences. Two possibilities are:

(a) Storage in the delay tanks for the low activity effluent. Here any leaching would contribute little to the activity already present in the effluent and the depth of water in the delay tanks would provide some shielding from the intense gamma radiation arising in the fixed bed.

(b) Burial on a suitable site would provide adequate shielding, and, provided the surrounding soil was impermeable, the leaching process would be slow. Clay soil would be most suitable since the clay itself would act as a cation exchanger and would suppress extensive diffusion of any leached fission products.

An interesting method of storage, combining burial and ion exchange has been applied for the last ten years at the Hanford Works,

U.S.A.   Here the medium and low activity wastes are allowed to percolate through the subsoil, where ion exchange takes place *in situ*. Obviously the method can only be applied under very favourable cirumstances, the requirements being low rainfall, deep ground-water level and suitable sub-soils.   It is unlikely that sites suitable for this means of disposal will be found in the United Kingdom.

REMOVAL OF STRONTIUM AND CAESIUM

Strontium-90 and caesium-137 are two of the most long-lived fission products, and both have a considerable potential value as radiation sources.   Strontium-90 is a pure beta emitter, with no gamma emission, and is therefore a useful source of beta particles for industrial purposes, since heavy shielding is not required. Caesium-137 also emits only beta particles of energy 1·2 MeV, but freshly separated Cs^{137} soon reaches radiochemical equilibrium with its daughter, 2·6 minute Ba^{137}, which emits gamma radiation of energy 0·6 MeV.   The gamma energy is such that a Cs^{137} source could replace Co^{60} in many applications.   Cobalt-60 is very much in demand and large units (tens or hundreds of curies) are often required.   At present Co^{60} is prepared by the expensive prolonged irradiation of natural cobalt in a nuclear reactor, and utilisation of Cs^{137} should bring about a considerable saving of valuable irradiation facilities.   During the year 1957–8 about 9500 curies of Cs^{137} were separated from the fission products at the Windscale plant and made up into eight gamma irradiation sources.

A further incentive to study methods of separating caesium and strontium from the rest of the fission products is provided by their high biological hazard.   The effect on the biological hazard of removing these species from a fission product waste is illustrated in Fig. 25.   This shows how rapidly the biological hazard falls off with time after the complete removal of strontium and caesium.   After a matter of fifteen years the bulk of fission products is then sufficiently innocuous to be regarded as low activity wastes and dealt with accordingly.   This implies that a smaller liquid storage capacity would be required because of the increased turnover of the stored liquors.   Furthermore, after a fifteen-year decay much of the material present is in the form of stable nuclides, since many of the fission-product chains involve half-lives of only a year or less.   These stable nuclides include such materials as ruthenium, rhodium, palladium and xenon, the total value of which has been estimated to be about £40,000 per annum at present market prices, for a programme involving the generation of 6000 MW of electrical power.

The situation is complicated, however, by the need for a very high percentage removal of caesium and strontium. For example, as may be seen from Fig. 25, if only 0·1 per cent of the original Sr^{90} and Cs^{137} were not removed the resulting hazard of the fission

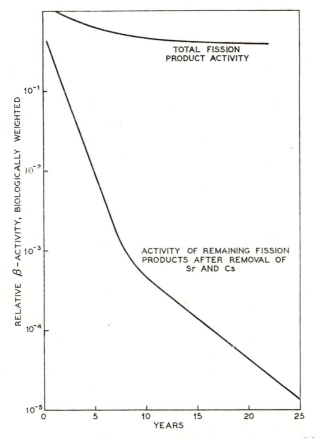

FIG. 25. FALL IN BIOLOGICAL HAZARD OF FISSION PRODUCTS WITH AND WITHOUT REMOVAL OF Sr AND Cs

products would be about five times greater than if all the Sr^{90} and Cs^{137} were removed. Furthermore after removal of Sr^{90} and Cs^{137}, alpha emitters such as Am^{241}, Pu^{238} and Cm^{242}, which are formed by multiple neutron capture in the reactor fuel, may contribute significantly to the biological hazard of the residual mixture. These materials may, in fact, set a limit to the degree which it is worth while to decontaminate the solution from caesium and strontium.

Several methods of separating caesium and strontium from the bulk of the fission products have been tried on a laboratory scale, while some have been applied on the pilot plant scale.

(a) *Hydroxide Precipitation.* Increasing the pH of a waste solution containing nitric acid results in the precipitation of all the species present in the solution with the exception of Sr and Cs. Several methods of achieving this have been described. Addition of ammonia to give a predetermined pH does not give a clean separation. At low pH ($\sim$ 6) very little precipitation occurs. At pH 8 most hydroxides are precipitated, but co-precipitate about 25 per cent of the strontium. Increasing the pH electrolytically results in a somewhat better separation. However, for the purposes in hand a much greater degree of removal (or *decontamination factor*) of Sr and Cs is required; this can be better achieved by direct precipitation of these species from solution.

(b) *Precipitation of Strontium.* Strontium nitrate is only slightly soluble in strong nitric acid, the solubility being further decreased by the presence of ferric nitrate, as illustrated by the results in Table 28.

TABLE 28.—SOLUBILITY OF STRONTIUM IN NITRIC ACID
AT 25°C

(*From U.N. Geneva Conference, Paper* 415, 1955, *by E. Gluekauf and T. V. Healy*)

Per cent HNO_3	99	90	80	70
Mg Sr/litre (no Fe) Mg Sr/litre (16 g Fe/l)	1·3	1·0 1·0	3·2 1·0	$\sim$ 20 5

The use of lead nitrate as a carrier also increases the removal of strontium, the lead being readily removed from the strontium at a later stage. The addition of nitric acid to a concentrated fission-product solution containing about 1 g/l of strontium results in the precipitation of 99 to 100 per cent of the latter. Further precipitation, with lead nitrate added, results in an overall decontamination of 10^4. A process based on nitric acid precipitation is at present used at the U.K.A.E.A. Radiochemical Centre, Amersham.

(c) *Precipitation of Caesium.* Several insoluble salts of caesium are known. Sodium phosphotungstate precipitates the caesium salt quantitatively in strong nitric acid solutions, but a large excess of reagent is required. The excess could be recovered for further use by solvent extraction, since it is soluble in many organic solvents. This precipitation is used at Amersham for Cs^{137} separation.

Caesium may also be precipitated as an alum by the addition of a large excess of potash alum, as tetraphenyl boron caesium by the addition of the sodium salt and as nickel caesium ferrocyanide by the addition of nickel ferrocyanide. None of the precipitations are particularly suitable for fission-product wastes, since a pH of between two and four is required, and at these low acidities impurities in the solution and some of the other fission products are precipitated.

Although the problems presented by the production of large quantities of fission products in power reactors are formidable, methods are potentially available for dealing with the output; further effort is, however, required to develop fully the more promising of these methods. One feature which considerably eases the disposal problem is the high degree of localisation of the radioactivity; the Windscale Works are expected to process the whole of the output of irradiated fuel elements from the United Kingdom nuclear power stations for many years to come. Considerable amounts of radioactive materials are being distributed for use in industry, research laboratories and hospitals. The materials which are supplied are, in general, either in a safe form (for example, Cs^{137}, Sr^{90} or Co^{60} radiation sources in sealed capsules) or as low-activity sources of comparatively short half-life (for example, I^{131} or Na^{24} for tracer studies or radiotherapy). Because of their rapid decay these isotopes constitute only a comparatively small ingestion hazard, and, as storage for several half-lives reduces the activity to negligible proportions, they constitute only a minor disposal problem.

FURTHER READING

Farmer. " Effluent Disposal at Windscale ", *Journal of the British Nuclear Energy Conference,* January 1957.

Saddington and Templeton. *Disposal of Radioactive Waste.* George Newnes Ltd. 1958.

United Nations. Geneva Conference, 1955. Vol. 9, *Reactor Technology and Chemical Processing.*

CHAPTER II

SOME FUTURE APPLICATIONS

THE primary aim of the atomic energy industry in its early years was the production of plutonium for military purposes. The conditions of operation of the nuclear reactors, such as those at Windscale, were optimised for plutonium production and the major chemical problems arose out of the processes required for the separation of this element from the parent uranium and from the fission products.

More recently, nuclear reactor designs have been optimised in favour of the production of heat energy for electric power generation. Extension of this aim over the next two decades will require many improvements in the performance of particular components of the reactors, such as pressure vessels, new metals for canning the fuel elements and possibly new coolant gases. New types of reactor may be favoured also, perhaps reactors of the homogeneous class employing aqueous solutions, liquid metals or fused salts. These will introduce many of the novel chemical problems which have been outlined in the preceding chapters. But what next?

There are several interesting long-term possibilities which are beginning to attract considerable attention. The first is the use of the energy which appears as beta and gamma radiation in spent fuel elements removed from nuclear reactors. It is shown in Table 5 that about 5 per cent of the total energy released by the fission process appears in this form. Up to the present time this energy has been entirely wasted, the fission products being separated from the uranium fuel and stored for long periods in relatively dilute solution.

Further possibilities are the utilisation of nuclear reactor energy directly, without the intermediate generation of electricity, either for process heat or to induce novel chemical reactions of industrial importance. The present position on each of these developments will be discussed in this chapter.

UTILISATION OF RADIATION FROM FISSION PRODUCTS

A small fraction of the theoretical total energy released in an operating nuclear reactor is stored in the radioactive fission products.

The latter consist of isotopes of elements from zinc to dysprosium with a wide variety of half-lives. The radiation energy from the mixed fission products is thus emitted over a considerable period of time after removal of spent fuel elements from a reactor, but only a fraction of it may be utilised in industrial processes since the latter require relatively short irradiation times in order to attain a worthwhile rate of production.

An approximate indication of the total quantity of radiation which might be available from fission products is given by the expectation that a 300 MW nuclear station, such as the one under construction at Bradwell, would release about 3 MW of heat as delayed beta and gamma radiation from the fuel elements, a few days after removal from the reactor. It would not, however, be very practicable to utilise this energy as heat either for manufacturing processes or for general space-heating because of the relatively rapid rate at which the activity dies away. New potential uses for fission product sources in medicine, agriculture and industry are continually being discovered. We do not propose to deal with this extensive subject here, but to show how the utilisation of this nuclear waste energy represents a more direct challenge to the skill of the chemist in developing chemical or physical effects which are specifically induced by radiation.

The manner in which fission product energy decays is illustrated in Fig. 26. It is evident that if the fission product activity can be used after only a short delay then a considerable increase in available radiation is achieved. However, the activity varies rapidly with time initially and this would lead to operational difficulties in applying uniform dose rates to the irradiated material. It may be advantageous, therefore, to allow considerable decay to take place before attempting to use this source of energy; the quantity available towards the end of the curve in Fig. 26 will be much less than at the beginning, but it will be relatively more constant.

The homogeneous aqueous reactor offers the possibility of working right at the top of the fission product decay curve. About one-fifth of the fission product decay energy (that is, about 1 per cent of the total energy of fission) is emitted from the gaseous products xenon and krypton during the first minute after each fission. The homogeneous reactor is designed specifically to reject these gases rapidly from the reactor circuit. If, however, the circuit characteristics were such that the gases took more than one minute to be removed into an irradiation device, the radiation available would be reduced to a maximum of about one-quarter of 1 per cent of the total fission energy.

A special difficulty would arise in this type of system from the fact that the very short half-life isotopes emit gamma radiation

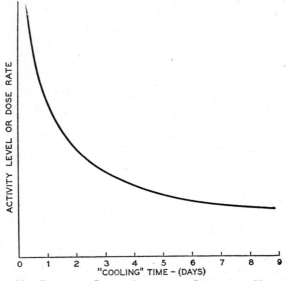

FIG. 26. DECAY OF GAMMA ACTIVITY OF IRRADIATED URANIUM

exceeding the 1 MeV threshold for photoneutron production from certain light atomic nuclei. This might lead to embarrassing induced activity in the chemical product and would also require the use of neutron shielding around the irradiation plant.

Most assessments which have been made to date on the utilisation of fission-product decay energy for industrial processes have considered three main types of irradiation unit:

(a) An array of spent fuel elements surrounded by, or interspersed with, pipes carrying the desired chemical reactants.

With the present types of fuel element processing for the separation of fission products and plutonium from uranium, it is necessary to allow irradiated elements to decay for periods of several months before the radioactivity is low enough to handle. Suggestions have been made that the radiation emitted during this decay period could be utilised by passing material for irradiation between an array of fuel elements. Although at this stage the radiation source may be considered to be in a relatively compact form and not too difficult to handle, there are several disadvantages. If the fuel elements are based on natural uranium, about two-thirds of the gamma energy and almost all of the beta energy will be absorbed in the elements and will not be available for use. A considerable amount of engineering thought would be required also to devise a means of moving the fuel elements around in order to keep the dose rates in the chemical reaction channels reasonably constant despite the fission product decay.

(b) Concentration of the mixed fission products which have been separated from the parent uranium into compact, intense sources of mixed radiation.

It is possible that this method may produce compact sources of reasonably constant radiation strength which would not be too difficult to transport from the separation plant to a chemical factory.

(c) Separation of particular fission products to give concentrated sources with an accurately known energy spectrum.

This facilitates the control of dose rate in the irradiated materials. Separated beta emitters would have the advantages of low shielding requirements and more efficient energy absorption in the irradiated system. The small range of beta particles would enable the irradiation unit to be quite compact (the range of 1 MeV beta particles in water is about 5 millimetres), but it would impose severe restrictions on the thickness of partitions between the source and the irradiated material. These partitions could not be more than a few thousandths of an inch thick.

The number of beta emitters of relatively long half-life and with reasonably energetic beta rays is rather limited: Y^{90} obtained as a short-lived daughter from long lived Sr^{90}, Pr^{144}, from Ce^{144}, and Rh^{106} from Ru^{106}. Suggestions have been made that the difficulty of the separating walls may be overcome by the use of Kr^{85}; since this is a rare gas it will be chemically unreactive and could be mixed homogeneously with the liquid system to be irradiated, and completely separated again later.

The increased range of gamma emitters at once eliminates the necessity to have very thin separating walls between the irradiation source and the chemical reaction system, but leads to problems of considerable shielding and possible dilution of the chemical product as a result of working with large volumes of solution in order to obtain reasonable energy absorption efficiency. The isotope Cs^{137} is being extracted on an increasing scale from fission product waste; its short-lived daughter, Ba^{137}, emits a penetrating gamma ray which is not energetic enough to produce undesirable neutrons in irradiated materials, and it has a reasonably long half-life of thirty years.

The essential differences between the utilisation of beta and gamma emitting sources are illustrated in Fig. 27.

Having seen how much energy is available from fission products, and some possible arrangements which have been suggested for

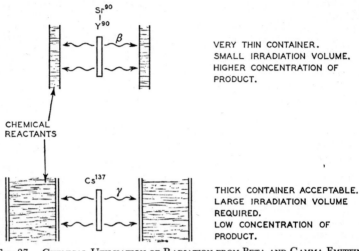

FIG. 27.   CHEMICAL UTILISATION OF RADIATION FROM BETA AND GAMMA EMITTING SOURCES

making use of this energy, we may now examine the type of system which may be irradiated.

RADIATION-INDUCED CHEMICAL REACTIONS

The heat equivalent of radiation is very low, for example the absorption of all the radiation from 1000 curies of 1 MeV gamma energy generates only about 6 watts. It is necessary, therefore, to examine what properties of radiation can be used in large-scale processes.

One property which is receiving considerable attention is the ability of radiation to destroy micro-organisms without an appreciable temperature rise in the irradiated material. Sterilisation of food, drugs, and other systems, is being extensively investigated; the Proceedings of a Conference on Nuclear Engineering held at the University of California in 1955 contain an estimate that it would be economical to build a nuclear reactor in the 3 to 60 MW range for the sole purpose of grain disinfestation or fresh meat pasteurisation. However, it is not proposed to describe these interesting possibilities in detail here, but to concentrate on developments of a more chemical nature.

If we wish to get worthwhile yields of chemicals for the expenditure of the rather limited sources of fission product decay energy, then we have to choose chemical reactions of the chain type such that radiation could supply the small amount of energy to start the reaction, and chemical energy would carry it the rest of the way. Polymerisation reactions appear to be very suitable, with the production of such widely used materials as butyl rubber, polyvinyl acetate and polyethylene. It is fortunate that one of the results of the interaction of ionising radiation with matter is the production of free radicals and these may be such as to initiate the polymerisations. We must be careful, however, that we do not just attempt to replace a relatively cheap polymerisation catalyst in the conventional process by a more expensive radiation source. The radiation chemist must show that advantages may be gained in terms of perhaps lower operating pressures, or temperatures, or a product with more desirable properties. Also it must be remembered that high-voltage machines are capable of producing the same kinds of radiation as those arising from fission product sources and they have the advantages of a constant and easily controllable power output.

The polymerisation of ethylene is an example where one might gain the advantage of lower operating pressures (and hence lower costs) by the use of radiation. Until quite recently polyethylene production has been performed under very high pressures, whereas preliminary laboratory experiments have shown that the radiation-induced polymerisation proceeds at appreciable rates even at atmospheric pressure. One of the possible advantages of using gamma radiation would be its property of penetrating most materials to a very considerable depth. This may be important for *in situ* polymerisation in intricate mould patterns.

It is thought that graft co-polymerisation under radiation may produce entirely new types of polymers, and preliminary tests are said to have been quite promising. The aim would be to combine two types of polymer molecules in an ordered way so as to give a product with the desirable properties of both. Thus, the grafting of vinyl carbazole to polyethylene could give better dielectric and high temperature properties, and the adhesive properties of polytetrafluorethylene may be improved by the surface grafting of styrene.

Apart from the possible use of radiation to induce polymerisation reactions, suggestions have also been made for the irradiation of polymers produced by the normal means. Cross-linking between polymer chains may be induced by irradiation of polyethylene, with the result that the poor mechanical strength of the unirradiated material above 100°C is considerably improved. Many new uses for polyethylene consequently become available, for instance as electric motor winding insulation and heat-sterilisable containers.

Whenever organic matter is subject to intense irradiation, some breakdown usually occurs. This was mentioned in Chapter 9 as one of the problems in the use of organic liquids as nuclear reactor coolant materials. Under some circumstances it may be possible to put this degradation to useful purposes. It has been shown that high-energy beta radiation can degrade wood, which is essentially highly polymerised glucose, to material which becomes digestible by the rumen bacteria which form the basis of the cow's digestive system. It would be of interest to examine whether the same effect could be obtained using spent reactor fuel elements.

The industrial use of fission product radiation for chemical purposes is mainly a matter of speculation at the present time. Experimental work of an exploratory nature on the effects of radiation on organic systems is being performed on an ever-increasing scale, but enormous fields of possibility remain as yet untapped. The cost of producing, transporting and operating concentrated fission-product sources of high specific activity on a really large scale is still rather uncertain. The chemist has an important role to play in both these aspects of the work; he must look for promising new chemical systems for irradiation, and he may have to devise methods such as have been discussed in the previous chapter for the separation of individual fission products from the large number which occur in irradiated uranium.

NUCLEAR REACTORS FOR PROCESS HEAT

One of the characteristics of the more advanced designs of nuclear reactor is their potentiality as sources of very high grade heat. The High-Temperature Gas-Cooled Reactor discussed in Chapter 8 is being designed to operate with a gas outlet temperature in the region of 800°C; this represents the upper limit to present large-scale design studies, but somewhat higher temperatures may be achieved at a later date.

There should be considerable uses for nuclear high-grade heat if it becomes available at a cost competitive with conventional energy sources. The steel industry, for instance, uses very large quantities of heat; a fully integrated iron and steel plant requires of the order of 1500 MW, derived primarily from coal. Moreover, the total United Kingdom annual consumption of fuel for the manufacture of iron and steel amounts to the equivalent of over

30 million tons of coal. Temperatures in the steel plants are too high for the application of known reactor technology, being in the range 1200 to 1600°C, nevertheless the possible replacement of so much coal represents a long-term aim which should not be lightly cast aside.

Coming a little further down the temperature scale, the production of lime by the roasting of calcium carbonate requires a temperature about 1100°C, the heat being supplied from coal at the rate of about 1 MW per kiln. The production of water gas by the reaction of coke with steam preheated to about 900°C could also lead to a considerable saving in coke consumption.

Although a decision on the technological feasibility of the application of nuclear heat to chemical processes at temperatures somewhat below 1000°C may be possible after the operation of a High-Temperature Gas-Cooled Reactor Experiment, there is also an economic obstacle to be overcome. Because of the high capital cost of nuclear reactors, cheap heat could be produced only in large installations. If the desired thermal output of a nuclear station falls below about 100 MW the cost of heat begins to increase rapidly. This is a considerable limitation, since very few chemical plants require heat from single units on such a large scale.

It may be necessary, in order to make use of the economy arising from increased scale, to look at new chemical processes rather than to try to adapt nuclear reactors to existing types of plant. There is, for instance, an enormous potential demand for cheap hydrogen for such purposes as coal hydrogenation. Investigations are being made at the A.E.R.E., Harwell, on the possibility of utilising nuclear heat to produce hydrogen by a modification of the steam-iron process. In the latter, steam is reduced to hydrogen by a lower oxide of iron which itself becomes oxidised.

Re-formation of the reduced oxide by heat alone requires excessively high temperatures, and a search is being made among oxides of the transition metals—such as manganese, vanadium and cobalt—in order to find a system in which the thermal reduction could proceed at a temperature about 800°C. Such a system, in which the oxide would be continuously cycled between the *oxidiser* and the *reducer*, would require only water as a feed material. Since the wasteful conversion of nuclear heat to electricity, and subsequent electricity transmission losses, would be eliminated, this process might well be more attractive than the electrolysis of water for hydrogen production.

The chemical industry must be somewhat conservative in outlook since new large-scale plants involve the investment of very large sums of money. Radical changes to existing technology cannot be considered lightly, and many years of preliminary work up to the pilot plant scale, involving experimental reactors, will be required before any particular process for using nuclear heat becomes accepted.

The extensive use of nuclear reactors as sources of relatively low grade heat appears to be possible on a much shorter time scale. The first use of nuclear energy for space heating was made as early as 1952: by-product heat from the experimental reactor BEPO was utilised for heating some of the adjacent buildings. There has been no further development along these lines in the United Kingdom, but a reactor is being designed in Sweden which will deliver 100 MW of heat in the form of steam at 120°C for general space heating in the city of Västerås. More pertinent to the subject matter of this book is the Norwegian 20 MW reactor at Halden which will be the first reactor to supply steam to an industrial plant. The steam will be supplied intermittently and on an experimental basis to a neighbouring pulp and paper plant. There may well be a future demand for reactors delivering 20 to 100 MW of heat for this purpose, particularly since a considerable number of paper mills are sited in areas where the cost of conventional fuels is high.

Some thought has been given to the possibility of using low-grade nuclear heat for the de-salting of brackish or saline water by evaporation techniques. Estimates show that such an application may be economically feasible on a very large scale, with daily fresh water outputs from a single unit of tens of millions of gallons. The provision of extensive fresh water supplies has not been an urgent problem for those countries which have pioneered the development of atomic energy; nevertheless, future application to this purpose may open up vast new areas in previously undeveloped countries for the spread of both agricultural and industrial activities.

THE DIRECT UTILISATION OF NUCLEAR REACTOR ENERGY FOR CHEMICAL SYNTHESIS

In the previous section we have mentioned briefly the use of nuclear reactors as sources of process heat for the chemical and metallurgical industries. For such applications it is envisaged that an intermediate circulating fluid would be used to transfer the heat and that the chemical reactants would not require to be passed through the nuclear reactor core. It is of considerable interest also to examine the possibility of performing industrially important chemical reactions within a nuclear reactor.

The kinetic energy of the neutrons and the gamma radiation are relatively small fractions of the energy produced in an operating reactor; moreover, neutrons and gamma rays do not interact very strongly with matter. It seems more desirable, on the basis of efficient energy utilisation, to attempt to tap the high kinetic energy of the fission fragments at the moment of fission: this constitutes about 84 per cent of the total fission energy.

At the present time little is known of the chemical reactions which may be induced by fission fragments. They are heavy and highly charged particles, and so they interact very strongly with matter; furthermore, it is possible that for an equivalent energy absorption the yield of product might be considerably greater from bombardment by fission fragments than by any other reactor radiation.

The fission fragment energy produced in a heterogeneous nuclear reactor which employs metallic uranium fuel elements cannot be utilised for chemical purposes. The penetration range of fission fragments in metals is only of the order of 2×10^{-4} in. and hence they are completely stopped by the canning materials. Similarly, in a solution fuel reactor of the homogeneous aqueous type the range is increased only to about 8×10^{-4} in. and to obtain sufficient interactions between fission fragments and the desired chemical system the latter would have to become the reactor fluid in place of water. Reactions of dissolved gases would not be suitable since it would not be possible to achieve sufficient solubility to ensure that the majority of the energy was deposited in the dissolved gas rather than in the carrier liquid.

REACTOR DESIGN

If we wish to make direct use of fission fragment energy on a large scale we are driven to a new type of reactor system: namely, one in which the uranium fuel is sufficiently finely divided to allow the fission fragments to escape into the chemical reactants. The particles, for instance of uranium oxide or carbide, would have to be less than 10^{-4} in. diameter in order not to absorb practically all the fission fragment energy. There is no special difficulty in producing particles of this size; it is known that oxalates of heavy metals may be decomposed thermally to produce oxides with a high surface area and a particle diameter in this region.

The simplest manner in which the chemical production system and the nuclear fuel system could be combined is shown in Fig. 28. The nuclear fuel particles are fluidised by the gases which it is desired to react, and after leaving the reactor the mixture is passed through a heat exchanger before separating the solid and gas phases again in order to extract the chemical product. The importance of the heat exchanger is stressed since it seems very unlikely that it would be economical to build a reactor for chemical production alone—it will be necessary to generate some electricity

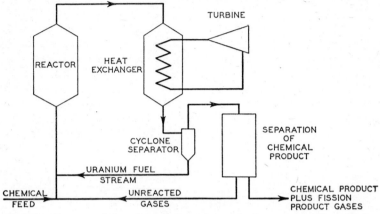

FIG. 28. REACTOR SYSTEM FOR DIRECT CHEMICAL PRODUCTION USING FISSION
FRAGMENT ENERGY

at the same time. We may illustrate this as follows:

Fraction of total reactor energy appearing as kinetic energy of fission fragments: 84 per cent.

Only about one quarter of this will be available for chemical purposes due to losses in the fuel particles and inefficient absorption in the gas system. The fraction of total reactor energy available then is reduced to 20 per cent.

Not all of this energy will be used to break chemical bonds: some will be lost in heating the chemical system. The maximum fraction of the total reactor energy used for direct chemical purposes consequently is reduced to: ~ 1.5 per cent.

This final figure is dependent upon the radiation chemical yield (G), and the value given was calculated for $G = 10$. This is probably a maximum attainable value. Conventional chemical manufacturing processes which derive their energy directly from coal, or oil, have higher thermal efficiencies, but the figure of 1·5 per cent may be comparable with some types of synthesis requiring the intermediate conversion of coal to electricity. Probably the closest analogy we may draw is with the Birkland-Eyde process for nitric acid manufacture. The efficiency of the electric arc in this process is about 5 per cent, and if the arc is fed with electricity generated either from coal or nuclear raw material the overall efficiency is reduced to about one third of this. Under such circumstances the arc process becomes completely uneconomical; it can be considered only if there is available an abundant supply of very cheap (hydro) electricity. Similarly, direct chemical production using fission fragment energy appears economically favourable only if the reactor operating conditions are chosen so that the remaining 98·5 per cent of reactor energy may be converted to electricity. Nevertheless, high rates of chemical production are potentially possible and at such time as the supply of coal or oil becomes more restricted, production from nuclear energy could become very desirable.

What sort of gases may be introduced to the reactor system? In order to cut wastage of neutrons to a minimum the gases must be of relatively low neutron cross-section. Elements which are acceptable in large quantities comprise quite a short list, but it is fortunate that several industrially important gases may be included as combinations of these elements.

Elements Acceptable in Large Quantities	Neutron Absorption Cross-section (barns)
O	$<0·0002$
D	$0·0006$
C	$0·003$
F	$0·009$
Be	$0·010$
Bi	$0·033$
Mg	$0·063$
$\left\{\begin{array}{l} \text{H} \\ \text{N} \end{array}\right.$	$\left.\begin{array}{l} 0·33 \\ 1·9 \end{array}\right\}$

Although the neutron cross-sections of hydrogen and nitrogen are relatively high, they are included as acceptable because, being gases, the amount present in the reactor is small even at the highest pressures considered. Furthermore, the product of a neutron capture in hydrogen is inactive, low cross-section deuterium. Neutron capture in nitrogen is more serious since it gives rise to the biologically hazardous C^{14} isotope by an (n,p) reaction. However, this may be neglected compared with the fission product Sr^{90} produced from nuclear fuel.

The use of gases requires a consideration of the limitations imposed by maximum operating pressures in the reactor system

If we consider utilising the whole of the reactor system for chemical production, then the present gas-cooled reactor designs using natural uranium fuel show that the massive size of the steel containing vessel imposes a restriction of the pressure to less than 10 atmospheres. On the other hand a fuel system highly enriched in U^{235} would need a much smaller pressure vessel and the maximum permissible pressure may be 100 atmospheres.

One may postulate two methods by which gaseous reactions may be promoted under fission-fragment bombardment: either the reactions proceed under the direct stimulus of energy absorption in the gas phase reactants, or energy absorbed in the solid fuel particles modifies their surface so that they become very efficient catalysts, and the reactions proceed under their influence. Present knowledge does not allow a decision to be made between these two mechanisms, and in any case the answer may be different for specific systems. The two mechanisms may lead to quite different reactor concepts, however.

For direct promotion of the chemical reaction by energy absorption in the gas phase, the efficiency will depend on the gas pressure: this should be as high as possible. Moreover, in a natural-uranium fuelled reactor, a dense suspension of particles will be required and most of the energy of the fission fragments will be lost by absorption in neighbouring particles and dissipation as heat in U^{238}. These two considerations then lead to the conclusion that only a highly enriched fuel and pressures above 10 atmospheres will be feasible for chemical applications. If, however, energy absorption in the solid phase is the controlling factor, there is no such restriction and a natural-uranium fuelled reactor would probably be advantageous since the fuel surface area would be higher.

The heavy-water moderated gas-cooled reactor might be a suitable type for adaptation to chemical production. The heavy water is contained in a large tank and is kept below 100°C by thermal insulation around the fuel element tubes. These pierce the tank and contain the fuel elements which are cooled by circulating gas. For chemical production, the conventional fuel elements would be replaced by small-diameter oxide particles fluidised by the chemical reactant gases. The high pressure of the latter would be contained by the thick walls of the fuel tubes, a more satisfactory arrangement than containment by a large pressure vessel surrounding the whole reactor core.

Radioactivity of the Product

We have stressed that it is important to choose gases of low neutron

cross-section for this application, and there will probably be negligible activity produced in this way.   However, the use of fuel
particles so small that the maximum fission-fragment energy escapes
and is transferred to the reactant gases implies that when the fission
products have lost their kinetic energy and may be considered as
normal elements or compounds they will be residing primarily in the
gas phase.   It is possible to estimate how great the radioactivity
would be and the order of decontamination factors required to make
the product acceptable.   In the specific case of the fixation of
atmospheric nitrogen (see below) and conversion to agricultural
fertiliser, the decontamination factor dictated by the Sr^{90} activity
level would need to be of the order of 10^8.   Some of this decontamination would be obtained by re-adsorption of fission products
on to the uranium fuel particles, and a further large step could be
obtained by distillation.   Certainly it does not seem impossible to
achieve the desired removal of radioactivity.

The Cost and Scale of Production

Two of the important factors which will govern whether a
particular chemical should be produced in a nuclear reactor
rather than by the conventional means are the rate of production
and the economics of the process.

It is not possible on present evidence to arrive at precise figures for
either of these factors, owing to the difficulty of extrapolating the
present cost and demand of the chemical products and because
present-day reactor designs and breakdowns of cost have been
optimised for electricity rather than chemical production.   Approximate figures can be derived for both cost and rate of production,
however;   these are of use in assessing the orders of magnitude
involved and the types of reaction which would repay further study.

The cost of electricity or chemical production will be sensitive to
the size of the reactor.   It is unlikely that a chemical manufacturer
would be interested in a reactor producing more than about 50 MW
of heat energy, whereas much of the improved economy of future
nuclear power stations is likely to result from their construction
in much larger units, perhaps 1500 MW.   To a first approximation
this size disadvantage factor of the chemical production reactor
can be taken into account by the use of the detailed breakdown of
cost for early, rather inefficient, gas-cooled reactors to be built in
the U.K., and assuming that these figures would apply in about
1975 to 1980.   This is the period when such chemical producing
reactors are likely to become of interest.   Since there are no published figures for the heavy-water moderated gas-cooled reactor,

we have to make the further (reasonable) assumption that the partial breakdown of cost is broadly the same as for a graphite-moderated reactor.

The approximate cost of power from the early gas-cooled reactors is shown in the second column of Table 29. The third column shows each item adjusted approximately for simultaneous chemical and electricity production. Factors which have been taken into account are an increased nuclear fuel investment due to circulation of the fuel around the heat exchanger circuit, increased shielding around the latter and the introduction of remote maintenance, and additional plant for separation of the chemical product (assumed to be 5 per cent of the total capital cost). It has been stated above that

TABLE 29.—APPROXIMATE COST OF POWER

	d/kWh Heat (Early Gas-cooled Stations)	d/kWh (Heat) with Simultaneous Chemical Production
Capital charges	0·12	0·13
Fuel inventory charges	0·02	0·16
Fuel consumption	0·08	0·16
Operation	0·01	0·015
	0·23	0·46

efficient utilisation of fission-fragment energy will probably require highly enriched fuel; the figures in the third column of the Table assume that plutonium will be used at a cost of £5000/kg.

If the reactor is designed to operate so that full electrical credit may be obtained, then the heat costs directly attributable to chemical production will be the difference between the totals of columns 3 and 2: namely, 0·23 d/kWh. The cost per short ton of any chemical product may then be obtained by substituting this value in the relationship:

$$\text{Cost/ton} = £12,100 \frac{(d/kWh)}{Gx \text{ (molecular weight)}}$$

G is the normal radiation chemical yield and (x) is the geometrica efficiency factor for energy utilisation; (x) will be about 0·2 to 0·3 in a system of highly enriched fuel particles fluidised by the chemical reactant gases.

By comparing the calculated figures with the cost and rate of production by conventional methods it is possible to predict the sort of values for G which will be required to make any particular reaction attractive.   For instance, if it appears that a reaction would be economically competitive when (Gx) is unity, this implies that fractional energy deposition of 20 per cent would require a G of 5, or if only 10 per cent deposition is possible then G must be about 10.

The rate of production of any chemical by fission-fragment bombardment in a reactor is given by

$$6 \cdot 6 \times 10^{-3} . QMGx \text{ tons per day}$$

where Q is the reactor power (MW heat) and M is the chemical molecular weight.   Large reactors are advantageous not only to increase the chemical throughput, but also because in general they have a better economy.

Some Suggested Chemical Reactions

The use of fission-product decay energy for chemical purposes, as discussed in the early parts of this chapter, is limited primarily to exothermic reactions.   There is no such restriction for fission-fragment energy utilisation since there is then an enormous amount of energy potentially available.

Development of laboratory investigations seems reasonably certain to produce chemical systems with many desirable products, but on the extremely limited evidence available at the present time it is feasible only to consider the less complex reactions amongst readily available gaseous starting materials.   A few of the industrially important reactions of this type are discussed below.   We should emphasise, however, that future applications in this field will most probably be along the—at present   completely unknown lines which may produce chemicals which are difficult or impossible to make in any other way.

Production of Nitric Acid

Early work in the BEPO reactor at Harwell showed that irradiation of air-water mixtures in a fast neutron-gamma flux gave a yield of 1·1 nitrate ions per 100 eV of energy deposited in the vapour phase.   At high doses, the decomposition of water gave hydrogen which reacted with further nitrogen to give detectable amounts of ammonium ions.   In 1956 it was reported from the U.S.A. that extension of this work to a study of nitrogen-oxygen mixtures at higher temperatures and pressures gave reasonable yields of nitrogen peroxide.   The G-value was as high as five molecules per 100 eV at 25 atmospheres pressure and 175°C.   The optimum conditions were said to be about 200°C, pressures in excess of 10 atmospheres, and a gas composition identical with that of air! It is unlikely that any other reaction will be as simple as this one, since it is necessary only to compress the air and drive it through the reactor. Other reactions will have appreciable raw material costs. The G-value is not known for fission-fragment bombardment at high temperature and pressure, but preliminary work at Harwell has shown that at normal pressure it is similar to that for neutron plus gamma irradiation.

The cost of manufacturing nitric acid by this method would be about £660 per ton for a 50 MW reactor if no credit is taken for reactor heat not utilised in the chemical reaction, or £44 per ton if we allow credit for electricity production; both figures assuming that $G \sim 5$ and $x \sim 0.2$. The present manufacturers' selling price of nitric acid is about £31 per ton. Evidently it will be necessary to determine whether there are any conditions under which the factor G will be considerably greater than 5 in order to make the product competitive with the conventional process.

It would require only four 500 MW nuclear reactors to produce the whole of the United Kingdom requirements (276,000 tons in 1951) if $Gx \sim 1$.

Production of Ammonia

There are several well-known processes for the fixation of nitrogen by reaction with hydrogen over a catalyst. The normal temperature of operation is about 500°C and pressures are very high. If this reaction could be made to proceed at much lower pressures under nuclear reactor conditions, it may become economically competitive due to the saving on the capital and maintenance costs of the very high pressure equipment.

Static irradiations of nitrogen-hydrogen systems at atmospheric pressure in the BEPO reactor have shown that $G \sim 0.7$ molecules of ammonia produced per 100 eV of energy deposited from mixed neutrons and gamma rays. It is to be expected that increase of pressure and irradiation by fission recoil fragments would both tend to increase the G-value. Assuming very optimistically that $Gx \sim 1$, the production cost of ammonia in a 50 MW reactor would be in the region of £164 per ton (neglecting the cost of the hydrogen raw material) compared with the present day market price of £52 per ton. There appears to be little hope of competing with the conventional process for this reaction. Moreover, the reaction does not appear as favourable as nitric acid production when we consider the possible annual output. Due to the lower molecular weight of ammonia, the output would be reduced and, if $Gx \sim 1$, it would require twenty-four 500 MW nuclear reactors to equal the United Kingdom demand (477,000 tons in 1951).

Hydrazine Synthesis

The possible synthesis of anhydrous hydrazine from nitrogen and hydrogen is worth investigation since it presents a rather special case. The normal Raschig process of synthesis from chlorine and ammonia, followed by extensive distillation for separating the hydrazine from water, is very expensive. Consequently, it is possible that synthesis in a nuclear reactor would be economically competitive with quite low values of G.

Hydrazine represents an extremely concentrated form of chemical energy and it has been suggested as a rocket propellant. The drawback to the large-scale use of hydrazine for this purpose seems to be the high cost. Cheap hydrazine may also find very extensive uses in the agricultural insecticide and the synthetic fibre industries.

Several investigators have attempted more simple syntheses of hydrazine, for instance by ultra-violet irradiation of nitrogen-hydrogen mixtures. The yield has been found to increase in the photolysis of ammonia to produce hydrazine when the product gases are rapidly swept out of the reaction chamber. A similar approach would be required for synthesis under nuclear reactor conditions.

If the rather optimistic figure of $(Gx) = 1$ could be achieved, one 50 MW reactor would produce over 3,000 tons per year of hydrazine.

Hydrogen

There is an enormous potential demand for cheap hydrogen both for general industrial purposes and for the large-scale hydrogenation of coal. One of the features of the homogeneous aqueous reactor system described in Chapter 7 is the intense decomposition of water. In the absence of the cupric sulphate recombination catalyst the radiolytic gas mixture may be extracted from the reactor circuit. Since the fissile material is in aqueous solution, it is possible to achieve a much higher energy utilisation than in the gas-cooled reactor, in fact (x) has its maximum value of unity.

The decomposition of water represents one of the very few systems in which the G-value for fission-fragment energy has been measured experimentally. If we use this G-value of about 1·5 molecules of water decomposed per 100 eV of energy absorbed, it is possible to calculate that a 50 MW nuclear reactor would produce about 4×10^5 cu. ft of hydrogen per day. It would be necessary to separate this from the oxygen and fission-product gases evolved at the same time, and this would increase the complexity of the plant. The major obstacle in this system is concerned with neutron economy, however. The homogeneous aqueous reactor will produce cheap power if it is based on the low neutron cross-section heavy water, and if it is built as a large unit. In order to produce large quantities of hydrogen, it would be necessary to have a light-water system and this would seriously affect the operating cost owing to neutron losses in the hydrogen. Reduction in the size of the reactor to produce about 10 MW of electricity would lead to the cost of the latter being rather high, about 1 d./u.s.o., and the value of the hydrogen would be only about 7 per cent of this. Thus, although a high value of (x) is possible, the low value of G implies that the hydrogen would be a relatively small by-product in the production of electricity.

Non-Aqueous Liquid Systems

We have confined our attention in this brief discussion primarily to gaseous reactions. It is possible that fission-fragment energy may be utilised for chemical synthesis in organic liquid systems, although the techniques involved would be rather more complex. A patent application has been published recently in which it is claimed that reactor irradiation of a slurry of uranium dioxide in methanol gave a product containing ethylene glycol, and a slurry in a mixture of ethanol and hexane gave a variety of products which included butanediols, octanols and dodecanes. The investigation of such systems is very much in its infancy and may be expected to provide some interesting and useful possibilities in the future.

It is evident that a promising system for the direct conversion of nuclear heat to chemical energy has not yet been discovered, but very wide fields of investigation remain to be covered.

OFF-PEAK ENERGY STORAGE

This chapter would not be complete without a word on some wider aspects of chemistry which may become intimately connected with the atomic energy industry for the purpose of energy storage. Present designs of nuclear power stations have a very high initial capital cost and the cost of the power produced is quite sensitive to the *load factor*. For this reason they have been considered primarily for use in large base-load stations which operate almost con-tinuously. In a densely populated country such as the United Kingdom, there should be little difficulty in maintaining many of the early nuclear stations in operation at a high load factor, variations in demand being met by smaller coal or oil-fired stations. The situation is more complex in sparsely populated areas, however, and to make nuclear stations economically feasible it may then be necessary to introduce some form of energy storage to cope with peaks in the power demand. Even in densely populated areas, energy storage will become desirable when the electrical energy derived from nuclear stations becomes a large fraction of the total output.

There are mechanical ways of achieving this storage, for instance pumping of water to high levels and compression of gases, but several chemical methods appear to merit consideration also. One of these is the use of a system of fuel cells and another would be the conversion of electricity to a form of chemical energy which if necessary could be transported in order to supply smaller and more remote requirements. Here again hydrazine production may be of interest. Off-peak operation of reactors in a densely populated and highly industrialised country might be more conveniently utilised to produce gaseous fuels, such as methane, which could be distributed either in a gas grid system or as a condensed liquid. Such chemical energy storage cannot, as yet, compete economically with potential energy storage for large centralised storage schemes, but may be feasible under other conditions such as cheap local raw materials for the chemical synthesis. The challenge is to devise a form of chemical energy storage by direct conversion from nuclear energy, eliminating the wasteful intermediate conversion to electricity—another facet of the general problem of chemical production discussed in the previous section.

FURTHER READING

Henley and Barr. "Ionising Radiation Applied to Chemical Processes and to Food and Drug Processing." *Advances in Chemical Engineering*, Vol. 1, (1956).
Nicholls. *Conference on the Utilisation of Heat from Nuclear Reactors*, U.K. Atomic Energy Authority Report, A.E.R.E. CE/R-2257 (1957).
Walton and Wright. *Symposium on the Utilisation of Radiation from Fission Products*, U.K. Atomic Energy Authority Report, A.E.R.E. C/R-1231 (1953).

INDEX